A Roadmap for the Uptake of Cyber-Physical Systems for Facilities Management

This is the first book to conceptualise and develop a roadmap for the adoption of cyber-physical systems (CPS) for facilities management (FM) in developing countries. It is argued that effective use of CPS can help to significantly improve issues such as extended processing time, poor data acquisition, ineffective coverage of facility maintenance history, and poor-quality control within the facilities management sector. Through a theoretical review of relevant technology adoption models and frameworks, *A Roadmap for the Uptake of Cyber-Physical Systems for Facilities Management* provides a clear insight into the required parameters for integrating CPS into facilities management.

The book will be beneficial to relevant stakeholders who face the responsibility of facilities and construction management, as it contributes to the growing demand for the adoption of digital technologies in the delivery and management of built infrastructure. Furthermore, it serves as a solid theoretical base for researchers and academics in the quest to expand the existing borderline on construction digitalisation, especially in the post-occupancy stage.

Matthew Ikuabe is a Post-Doctoral Research Fellow at the University of Johannesburg, South Africa.

Clinton Aigbavboa is a Professor in the Department of Construction Management and Quantity Surveying and Director of the CIDB Centre of Excellence & Sustainable Human Settlement and Construction Research Centre, the University of Johannesburg, South Africa.

Chimay Anumba is both Dean and Professor at the College of Design, Construction and Planning, the University of Florida, USA.

Ayodeji Oke is a Senior Lecturer in the Department of Quantity Surveying, Federal University of Technology, Akure, Nigeria, and a Research Fellow in the Department of Construction Management and Quantity Surveying, the University of Johannesburg, South Africa.

Routledge Research Collections for Construction in Developing Countries

Series Editors: Clinton Aigbavboa, Wellington Thwala, Chimay Anumba, David Edwards

A 21st Century Employability Skills Improvement Framework for the Construction Industry
John Aliu, Clinton Aigbavboa & Wellington D. Thwala

Construction Project Monitoring and Evaluation
An Integrated Approach
Callistus Tengan, Clinton Aigbavboa and Wellington D. Thwala

Developing the Competitive Advantage of Indigenous Construction Firms
Matthew Kwaw Somiah, Clinton Aigbavboa and Wellington D. Thwala

Construction Digitalisation
A Capability Maturity Model for Construction Organisations
Douglas Aghimien, Clinton Aigbavboa, Ayodeji Oke and Wellington D. Thwala

Sustainable Construction in the Era of the Fourth Industrial Revolution
Ayodeji E. Oke, Clinton O. Aigbavboa, Stephen S. Segun and Wellington D. Thwala

Construction in Indonesia: Looking Back and Moving Forward
Toong-Khuan Chan and Krishna Suryanto Pribadi

A Maintenance Management Framework for Municipal Buildings in Developing Economies
Babatunde Fatai Ogunbayo, Clinton Aigbavboa and Wellington D. Thwala

Moving the Construction Safety Climate Forward in Developing Countries
Sharon Jatau, Fidelis Emuze and John Smallwood

Unpacking the Decent Work Agenda in Construction Operations for Developing Countries
Tirivavi Moyo, Gerrit Crafford, Fidelis Emuze

An Integrated Infrastructure Delivery Model for Developing Economies: Planning and Delivery Management Attributes
Rembuluwani Bethuel Netshiswinzhe, Clinton Aigbavboa and Wellington D. Thwala

A Roadmap for the Uptake of Cyber-Physical Systems for Facilities Management
Matthew Ikuabe, Clinton Aigbavboa, Chimay Anumba and Ayodeji Oke

A Roadmap for the Uptake of Cyber-Physical Systems for Facilities Management

Matthew Ikuabe, Clinton Aigbavboa, Chimay Anumba and Ayodeji Oke

LONDON AND NEW YORK

First published 2023
by Routledge
4 Park Square, Milton Park, Abingdon, Oxon OX14 4RN

and by Routledge
605 Third Avenue, New York, NY 10158

Routledge is an imprint of the Taylor & Francis Group, an informa business

British Library Cataloguing-in-Publication Data
A catalogue record for this book is available from the British Library

Library of Congress Cataloging-in-Publication Data
Names: Ikuabe, Matthew, editor. | Aigbavboa, Clinton, editor. | Anumba, C. J. (Chimay J.), editor. | Oke, Ayodeji E., editor.
Title: A roadmap for the uptake of cyber-physical systems for facilities management / Matthew Ikuabe, Clinton Aigbavboa, Chimay Anumba, Ayodeji Oke.
Description: Abingdon, Oxon ; New York : Routledge, 2023. | Includes bibliographical references and index.
Identifiers: LCCN 2022061253 | ISBN 9781032446660 (hbk) | ISBN 9781032452791 (pbk) | ISBN 9781003376262 (ebk)
Subjects: LCSH: Facility management--Data processing. | Real estate management--Data processing. | Cooperating objects (Computer systems) | Commercial buildings--Maintenance and repair--Data processing. | Intelligent buildings.
Classification: LCC TS177 .R63 2023 | DDC 658.2--dc23/eng/20230130
LC record available at https://lccn.loc.gov/2022061253

ISBN: 978-1-032-44666-0 (hbk)
ISBN: 978-1-032-45279-1 (pbk)
ISBN: 978-1-003-37626-2 (ebk)

DOI: 10.1201/9781003376262

Typeset in Times New Roman
by Taylor & Francis Books

Contents

Illustrations

Figures

Tables

Preface

Facilities management (FM) has evolved over the years to become one of the cardinal bases for organisations to achieve their pre-determined objectives. This is due to the multi-disciplinary roles engaged in by facilities management activities to ensure a well-coordinated and synchronised delivery of mandates by the various units making up an organisation. However, the delivery of facilities management tasks is still plagued by inefficiencies and poor processes, because of the outdated techniques and approaches deployed in facilities management functions. Issues such as extended processing time, poor data acquisition, ineffective coverage of facility maintenance history, and poor quality control persist. In the current era of the Fourth Industrial Revolution, several sectors of the economy are fast gaining pace in integrating the emerging technologies, which have been proven to deliver the timely completion of tasks, cost optimisation, efficient service delivery, and improvements in health and safety, among others.

The focus of this book is on the development of a model for cyber-physical systems (CPS) adoption for facilities management. The book was conceptualised against the backdrop of the seeming challenges facing the engagement and delivery of facilities management, coupled with the glaring benefits to be accrued from the integration of technological innovations in its processes. CPSs are computing systems that function on the premise of the integration of computational capabilities and physical processes. They work through the monitoring of computer-based algorithms fused with the internet. They serve as the link between the physical and virtual worlds. The integration of CPS in built-up facilities would aid in fostering the seamless management of the facility, thus, encouraging real-time computation/monitoring of the state of the components making up a facility. To effectively incorporate CPS in FM, there is a need to clearly define a roadmap that would aid the espousal of the innovative technology.

Therefore, this book provides a clear insight into the required parameters for the integration of CPS in FM, with an emphasis on their use in developing economies. This is achieved by an adoption model, which was formed through the theoretical scrutiny of frameworks of adoption models for technological innovations. The highlight of the book is the presentation of the required parameters for the espousal of CPS for FM. This book will be of immense benefit to relevant stakeholders faced with the responsibility of facilities and construction management, as it contributes to the growing demand to integrate digital technologies into the delivery and management of the built infrastructure. Furthermore, the book serves as a solid theoretical base for researchers in the quest to expand the existing borderline on construction digitalisation, especially in the post-occupancy stage of construction projects.

Matthew Ikuabe
Clinton Aigbavboa
Chimay Anumba
Ayodeji Oke

Part I

Background

1 Introduction

Introduction

This book aims to develop a roadmap for the uptake of cyber-physical systems (CPS) for facilities management (FM). This chapter gives an insight into the background of the study while also attempting to reveal the need for the book and thereafter present the significant results from the outcome of the study discussed in the book.

Background and significance

Facilities management (FM) is a term encompassing a wide spectrum of services spanning the management of real estate, the maintenance of buildings, financial systems, the management of contracts, health and safety and domestic services (Atkin and Brooks, 2000; Amaratunga, et al., 2000). The South Africa Facilities Management Association (SAFMA) (2012: 5) defines FM "as a process that enables sustainable enterprise concerning the whole life management of workplaces to achieve productivity and support business effectively". Also, the International Facility Management Association (IFMA, 2019) outlines FM as a profession that comprises diverse disciplines and guarantees functionality of the built infrastructure by incorporating process, people, place, and technology. According to Kamaruzzaman and Zawawi (2010), FM occupies a strategic position to strike a balance between the management of technical processes or services and business concerns. Kok et al. (2011) stated that FM covers a range of service solutions: productivity, hospitality, safety, sustainability, and accessibility. The role of FM in the post-occupancy stage of the life-cycle of built infrastructures is increasingly becoming important, considering the strategic values that accompany it, since it contributes to the evaluation of the performance of infrastructures. Building performance measurement serves as a critical aspect to assess post-occupancy evaluation of the built infrastructure; hence the management of completed infrastructures contributes immensely to its value and overall performance. Chotipanich (2004) asserted that the function of FM is prominent in facility resource management, support services, inclusive of the working environment, and aiding organisations' business mandate.

DOI: 10.1201/9781003376262-2

To a considerable extent, the business environment determines the projected objectives of the organisation, thus implying that the functions of FM are laid out in preferential order to attain the organisation's core objectives. In a bid to consistently align an organisation's operations with the changing business environment, a review of the prioritised objectives of an organisation is constantly undertaken. Green and Jack (2004) noted that the harmonious integration of people, processes, and facilities improves the delivery of the organisation's integral duties, achieving the objectives, and enhancing the organisation's competitive advantage. FM serves as the connecting link between the physical space, the organisation, and its employees (Grimshaw, 1999). Carder (1997) refers to this link as "the interface between an organisation's core business and its physical working environment", likewise naming the facilities manager the "interface manager". The facilities manager is responsible tor developing, operating, and maintaining pre-determined standards and the built facilities and services to enable the right environment and suitable facilities that are most cost-effective for the employees' tasks and to attain the organisation's objectives. Considering the dynamic nature of the workplace, the need for in-depth research is imperative. Response to workplace change, strategic planning, setting, and culture are important, and facilities managers must be proactive (Kok et al., 2011). The results of such research ought to serve as a driver for the provision of tools that aid facilities managers to deal with varied situations and the ambiguity unique to the industry on which the research is based (Grimshaw, 1999). A feedback mechanism needs to be developed through a cordial relationship between the employees of an organisation and the facilities manager, as this would help in ascertaining performance through the measurement of users' satisfaction, in order to improve services, enhance delivery and meet the core mandate of the organisation.

The mandate of operational FM is targeted at organisational and administration processes. This ranges from activities such as controlling, monitoring, and managing operations in FM; policy, strategy and implementing plans, and ensuring that activities and processes are carried out, while adhering to the organisation's set standards and requirements. Facilities management entails a wide spectrum of operations and activities. These range from the management of space and property, to control of the environment, health and safety aspects, control and monitoring of various units of the organisation, and support services (Alexander, 2003). FM as a discipline covers a wide range of services that includes the management of real estate, contract management, change management, financial management, health and safety, and human resource management, in conjunction with the maintenance of a building, and other services, such as the supply of utilities, security and cleaning (Atkin and Brooks, 2000). Furthermore, Booty (2009) included the following in the functions of FM: planning of master space, finance budgeting and control, space inventory, construction management, design and layout, and maintenance management. Regarding the activities that come under the umbrella of FM, Collings (2007) stated they fall into the following areas:

cleaning, plumbing, energy management, pest control, electrical, landscaping, ventilation, tenant management, sustainability management, car parking, air-conditioning, and mailroom.

Other areas within the purview of FM are planned maintenance, restoration, and upgrades, as the key basis for the role of FM (Booty, 2009). Efforts are put into maintenance management so that the maintenance activities suitably accommodate any anticipated scheduled or unscheduled, preventive maintenance and repairs, in order to forestall any unexpected problems that will affect activities or the protection of lives and properties. Ihsan and Alshibani (2018) stated that issues of technical concern affecting the maintenance of facilities and personnel operations cover: updating IT services, maintenance of internal and external components of buildings, management of obsolete building components and keeping up with the increasing safety and environmental demands. Also, facilities managers must ensure they comply with statutory demands that control maintenance management.

In the current era of the Fourth Industrial Revolution (4IR), the increased application of information/digital technology is evident in various spheres of human endeavour. In recent times, basic human activities and industrial processes have experienced an upsurge in digital technologies. The McKinsey Global Institute (2016) observed that industries such as banking, manufacturing, health, and medical facilities have witnessed multidimensional access to digital technology adoption. According to Ikuabe et al. (2022), with the abundant development challenges facing twenty-first-century society and as detailed in the UN Sustainable Development Goals (SDGs), no people or profession can flourish without professional approaches, ideas, and goals based on the technological realm of knowledge. The built environment is equally not omitted from the scheme of things. The adoption of digital technology in construction processes has helped timely project delivery, the expansion of profit, delivery of quality, and increased customer satisfaction (Staub French and Khanzode, 2007). One of the distinct emerging technologies in various spheres of human endeavour is cyber-physical systems (CPS).

CPS are computing systems that function to integrate computational capabilities and physical processes. They are supervised by computer-based algorithms blended with the internet. They serve as the connection between the physical and virtual realms. CPS are also known as "realisation technology" (Kim and Park, 2013). The National Science Foundation (2013: 34) noted that CPS are "systems where physical and software components are deeply intertwined. Each operating on different spatial and temporal scales, exhibiting multiple and distinct behavioural modalities, and interacting with each other in a myriad of ways that change with context." Adaptation, intelligence, and responsiveness are achieved through the knowledge and information shared between the physical process and the computational decision components of CPS (Leitão et al., 2015). This is due to the fusion of computers and network monitors, which eventually monitor the physical procedures, with response loops where physical activities impact computations, and vice versa (Wang et al., 2012).

CPS are widely employed in several disciplines. Efforts have been made to navigate their application in various aspects of human life. One such effort is the CyPhERS ("Cyber-Physical European Roadmap and Strategy") (Cerngarle et al., 2013). In its 7th Framework Programme, the European Commission aimed at combining and developing the resources of the EU's mobile computing systems and strongly accessing them on networked embedded systems. The main aim is to build regional strategic research and innovation hubs motivated by CPS to safeguard the EU's competitiveness in this emergent field. Also, through the National Science Foundation, the US government financed a research-driven programme on CPS costing $150 million (R2 billion), placing CPS at the top of the preference list for federal research investment. Furthermore, nations such as China, Japan, and South Korea have stepped up research schemes to aid the placement of CPS on the map of innovative technologies (Kim and Park, 2013).

The use of CPS offers several benefits, depending on the field of application. The system provides the platform for the provision of solutions to problems in real time in the digital age (Bhrugubanda, 2015). Also, it enables the interrelationship between several technologies, such as control systems, real-time systems, distributed systems and wireless sensor networks. Herterich, Uebernickel and Brenner (2015) confirmed that CPS present opportunities for the reduction of downtime experienced with the use of the equipment and increase operation efficiency. The fusion of physical and computational processes in production operations brings about effective delivery, from the synergy of the dual process. Furthermore, innovations such as the product cloud, which serves as the domain for the storage and analysis of operational data, are significant determinants for the optimisation of the system. The information from the analysed data aids in the intentional planning and organisation of established models of production, for improved output and future system improvement. It helps the evolving system to reform and resize to achieve distinguished economic opportunities.

Recently, there has been an upsurge in the search for improved construction project delivery, coupled with infrastructure systems maintenance through virtual models (Popov et al., 2013). Virtual models are designed as a prototype of the envisaged built facility, and when the construction of the facility is completed, as-built drawings are designed to give a true reflection of the various components that make up the finished facility. Akanmu, Anumba and Messner (2013) noted that integrating CPS into the construction processes would potentially improve real-time monitoring of work progress, process control, and maintenance, updating alterations, exchange of information, and as-built documentation. The integration of CPS into built-up facilities in conjunction with the as-built virtual models would foster the seamless management of the facility. This set-up would encourage real-time computation of the state of the components in the facility. Based on the aforementioned premise, this study is geared towards developing an adoption model for the application of CPS in the management of built facilities. The anticipated model would guide the ease of implementation and usage of CPS in FM.

Objectives of the book

Organisations must ensure that the strategies employed for business success are in synchronisation with the projected FM objectives. However, due to outdated methodologies that have failed to meet the pace of the increasingly technological world, the synergy between optimally delivering FM activities and attaining business goals is missing. Therefore, there is a conscious need to re-navigate the approaches to the processes involved in meeting the mandates of FM. Against this backdrop, the study aims to develop a cyber-physical systems adoption model for facilities management to improve the infrastructural facilities' performance. Therefore, the following objectives were set by the study:

1 To identify the factors that determine the adoption of CPS in facilities management.
2 To identify the theories and models involved in technology adoption.
3 To establish the main and minor attributes that influence the adoption of CPS in facilities management.
4 To develop a cyber-physical systems adoption model for facilities management.

Contribution and value

CPSs have gained ground since the concept was first mentioned by Helen Gill in 2006 at the National Science Foundation in the United States. As earlier stated, quite a number of studies have been carried out on the application of different construction processes. However, there seems to be a lack of data concerning post-occupational operations of the construction life-cycle. The application of CPS in the management of facilities is an unexplored area that is covered by this research. Furthermore, the construction industry is still characterised by inefficiencies in processes, resulting from outdated approaches and methods of service delivery (Aghimien et al., 2021; Ikuabe and Oke, 2019; Ikuabe, Aghimien, Aigbavboa and Oke, 2020; Ikuabe, Aghimiem, Aigabvboa, Oke, and Ngaj, 2021; Ikuabe, Aghimiem, Aigabvboa and Oke, 2021). It cannot be denied that the management of construction projects at the pre-occupancy and post-occupancy stages needs a shift from conventional approaches to more innovative systems such as cyber-physical systems.

Facilities management entails a multi-disciplinary approach to the actualisation of the organisation's objectives. However, an organisation is often faced with challenges, such as keeping accurate records, cost control, lack of qualified personnel, and inadequate maintenance. The development of a model for the adoption of CPS that integrate computational and physical processes would aid in delivering optimised performance for facilities management tasks. Hence, the use of the system would aid in the real-time computation of the state of the components making up the facility. Also, if the components of the facility malfunction, bi-directional coordination between the physical component and the virtual model, synchronised by the computational ability of the CPS, sends notifications to prompt correctional measures (Akanmu et al., 2013; Ikuabe, Aghimien, Aigbavboa and Oke, 2020). With this in place, the goals of FM would easily be attained; and also, this would help solve the challenges encountered with the traditional model of FM.

Virtual models are designed as a prototype of the envisaged built facility. When the facility is completed, the as-built drawings are designed to reflect the various components that make up the finished facility. One primary goal for creating the as-built drawings is to facilitate easy and convenient management of the completed structure, with a prominent emphasis on the maintenance of the facility. The integration of CPS in the built facility in conjunction with the as-built virtual models fosters the seamless management of the facility. This set-up encourages real-time computation of the state of the components making up the facility.

The UN Sustainable Development Goal (SDG) number 9 seeks a built resilient infrastructure, promotes inclusive and sustainable industrialisation, and fosters innovation. A significant contribution of CPS is the efforts geared towards the actualisation of SDG 9 as mentioned earlier. If a holistic CPS adoption model for FM were established, the innovative technological advantage for the efficient and effective management of facilities would encourage the production of resilient infrastructure in the construction industry. Also, this would aid in fostering innovation, as stipulated by the SDGs. CPS serve as vehicles for delivering the mandates of the Fourth Industrial Revolution. Hence, their adoption for facilities management would create innovative methods and techniques to achieve the mandates of facilities management.

Structure of the book

Part I Background

Chapter 1, "Introduction", highlights the background of the research and the identified problem that seeks answers. The general description of the study is presented with the aim, research questions and objectives of the study. Also, the purpose, motivation and significance of the study are also highlighted in this chapter.

Part II Facilities management

Chapter 2, "Theoretical perspectives on facilities management", provides a theoretical review of FM that serves as a major aspect of the research. A literature review is presented in this chapter, focused on the history, definition, and functions of facilities management. Also, an assessment of building maintenance is conducted, while digital technologies for facilities management are discussed. Chapter 3, "Digital technologies for facilities management", provides an insight into the various digital technologies in the Fourth Industrial Revolution employed to deliver FM functions. Underpinned by the comparative advantage from the use of innovative technologies for FM delivery, the benefits associated with these technologies are revealed. Also, the challenges posed by the digitalisation of FM functions are discussed.

Part III Cyber-physical systems

Chapter 4, "Theoretical perspectives on cyber-physical systems", reviews the literature on cyber-physical systems. It gives an overview of CPS, outlining the requirements for their use, and projecting the designs and modelling of the systems. Also, an exposition of the use of CPS for facilities management is offered, and, finally, a bibliometric review of the study of cyber-physical systems in the construction industry was carried out.

Part IV Theoretical perspectives on technology adoption

Chapter 5, "Theoretical perspectives on technology adoption", presents a detailed review of the theories and models on technology adoption. A chronological and sequential detailing of these theories and models, based on how they have evolved over time, is presented. Also, the theoretical underpinnings for the study are outlined in this chapter. Chapter 6, "Gaps in technology adoption research", explores the gaps identified in the literature that prevent technology adoption. This reveals the gaps peculiar to adopting CPS for FM. The gaps identified are the business environment and performance measurement. These constructs were deemed crucial to the adoption of CPS for FM.

Part V The CPS adoption model for facilities management

Chapter 7, "Conceptual perspective of cyber-physical systems adoption model for facilities management", focuses on the conceptualised model for adopting cyber-physical systems for facilities management, which is a multidimensional framework with six constructs: (1) performance expectancies; (2) social influence; (3) enabling measures; (4) effort expectancies; (5) business environment; and (6) performance measurement. These variables are products of a well-executed literature review. Chapter 8, "Assessing the conceptualised cyber-physical systems adoption model for facilities management: a Delphi study", presents the results of the first and second rounds of the Delphi study undertaken in the research. The computations were carried out on the individual questions posed to the panel of experts, In contrast, the questions dwelt on the significance of the main and minor attributes in adopting cyber-physical systems for facilities management.

Conclusion

This chapter introduced the research by giving an overview of the main subject of the study. It highlights the background of the research and the identified problem that seeks an answer. Also, the general description of the study is presented with the research questions, and objectives of the study.

References

Aghimien, D.O., Ikuabe, M.O., Aigbavboa, C.O., and Shirinda, W. (2021). Unravelling the factors influencing construction organisations' intention to adopt big data analytics in South Africa. *Construction Economics and Building*, 213), 262–281.

Akanmu, A., Anumba, C., and Messner, J. (2013). Active monitoring and control of light fixtures during building construction and operation: Cyber-physical systems approach. *Journal of Architectural Engineering*, 20(2), 16–24.

Alexander, K. (2003). A strategy for facilities management. *Facilities*, 21(11), 269–274.

Amaratunga, D., Baldry, D., and Sarshar, M. (2000). Assessment of facilities management performance – what next? , *Facilities*, 18(1/2), 66–75.

Atkin, B. and Brooks, A. (2000). *Total Facilities Management*. Oxford: Blackwell Science.

Bhrugubanda, M. (2015). A review on applications of cyber physical systems. *International Journal of Innovative Science, Engineering and Technology*, 2(6), 728–730.

Booty, F. (2009). *Facilities Management Handbook*, 4th edn. Oxford: Butterworth-Heinemann.

Carder, P. (1997). The interface manager's toolkit. *Facilities*, 15(3/4), 84–89.

Cengarle, M., Bensalem, S., McDermid, J., Passerone, R., Sangiovanni-Vincentelli, A. and Törngren, M. (2013). CyPhERS – CyberPhysical European Roadmap & Strategy, 7th Framework Programme. Available at: http://www.cyphers.eu/sites/default/files/D2.1.pdf

Chotipanich, S. (2004). Positioning facility management. *Facilities*, 22(13), 364–372.

Collings, T. (2007). *Building and operations management*. Paper presented at the NZPI CPD Seminar, Wellington, NZ, April.

Green, A. and Jack, A. (2004). Creating stakeholder value by consistently aligning the support environment with stakeholders needs. *Facilities*, 22(13/14), 359–363.

Grimshaw, B. (1999). Facilities management: The wider implications of managing change. *Facilities*, 17(1/2), 24–30.

Hererich, M., Uebernickel, F. and Brenner, W. (2015). *The impact of cyber-physical systems on industrial services in manufacturing*. 7th Industrial Product-Service Systems Conference, Procedia CIRP 30, 323–328.

IFMA (International Facility Management Association) (2019). Definition of facility management. Available at: http://www.ifma.org/about/aboutifma/history#sthash.UAeyxW1Y.dpuf (accessed 3 November 2021).

Ihsan, B. and Alshibani, A. (2018). Factors affecting operation and maintenance cost of hotels. *Property Management*, 36(3), 296–313.

Ikuabe, M.O., Aghimien, D., Aigbavboa, C. and Oke, A.E. (2020). *Exploring the adoption of digital technology at the different phases of construction projects in South Africa*. Proceedings of the International Conference on Industrial Engineering and Operations Management, Dubai, UAE, 10–12 March, pp. 1553–1561.

Ikaube, M.O., Aghimiem, D., Aigabvboa, C. and Oke, A. (2021). Sustainable road infrastructure in rural areas in South Africa: A preliminary study. In R.J. Howelett, J.R. Littlewood and L.C. Jain (eds), *Emerging Research in Sustainable Energy and Buildings for a Low Carbon Future: Advances in Sustainability Science and Technology*. Cham: Springer, pp. 331–338.

Ikaube, M.O., Aghimiem, D., Aigabvboa, C., Oke, A. and Ngaj, Y. (2021). Barriers to the adoption of zero-carbon emissions in buildings: The South African narrative. In R.J. Howelett, J.R. Littlewood and L.C. Jain (eds), *Emerging Research in Sustainable Energy and Buildings for a Low Carbon Future: Advances in Sustainability Science and Technology*. Cham: Springer, pp. 135–148.

Ikuabe, M.O., Aigbavboa, C., Anumba, C., Oke, A. and Aghimein, L. (2022). Confirmatory factor analysis of performance measurement indicators determining the uptake of CPS for facilities management. *Buildings*, 12(4), 466.

Ikuabe, M.O., Aigbavboa, C. and Oke, A. (2020b). *Cyber-physical systems: matching up its application in the construction industry and other selected industries*. Proceedings of the International Conference on Industrial Engineering and Operations Management, Dubai, UAE, 10–12 March, pp. 1543–1552.

Ikuabe, M.O. and Oke, A. (2019). Contractors' opportunism: Construction professionals' awareness of influencing factors. *Journal of Engineering, Design and Technology*, 17(1), 102–114.

Kamaruzzaman, S.N. and Zawawi, E.M.A. (2010). Development of facilities management in Malaysia. *Journal of Facilities Management*, 8(1), 75–81.

Kim, K. and Park. K. (2013). An overview and some challenges in CPS. *Journal of the Indian Institute of Sciences*, 93, 1–8.

Kok, H.B., Mobach, M.P. and Omta, O.S. (2011). The added value of facility management in the educational environment. *Journal of Facilities Management*, 9(4), 249–265.

Leitão, P., Colombo, A. and Karnouskos, S. (2015). Industrial automation based on cyber-physical systems technologies: Prototype implementations and challenges. *Computers in Industry*, 81, 11–25.

McKinsey Global Institute. (2016). A route to higher productivity: Reinventing construction. Available at: https://www.mckinsey.com/~/media/McKinsey/Industries/Capital%20Projects%20and%20Infrastructure/Our%20Insights/Reinventing%20construction%20through%20a%20productivity%20revolution/MGI-Reinventing-construction-A-route-to-higher-productivity-Full-report.ashx (accessed 20 August 2021).

National Science Foundation (NSF) (2013). Cyber physical systems. Report no. NSF10515. Arlington, VA: NSF. Available at: http://www.nsf.gov/pubs/2010/nsf10515/nsf10515.htm (accessed 26 August 2021).

Popov, V., Juocevicius, V., Migilinskas, D., Ustinovichius, L. and Mikalauskas, S. (2013). The use of a virtual building design and construction model for developing an effective project concept in 5D environment. *Automation in Construction*, 19(3), 357–367.

South African Facilities Management Association (SAFMA) (2012). What is FM? Available at: http://www.safma.co.za/portals/0/What_is_FM_presentation.pdf (accessed 28 October 2021).

Staub-French, S. and Khanzode, A. (2007). 3D and 4D modeling for design and construction coordination: Issues and lessons learned. *ITcon*, 12, 381–407.

Wang, P., Xiang, Y., and Zhang, S. (2012). *Cyber-physical system components composition analysis and formal verification based on service-oriented architecture*. 9th IEEE International Conference on e-Business Engineering, pp. 327–332.

Part II

Facilities management

2 Theoretical perspectives on facilities management

Introduction

Facilities management (FM) is transdisciplinary or multidisciplinary, as it makes use of the knowledge from design, engineering, architecture, management, accounting, finance, and behavioural science (Teicholz, 2001). Hence, considering FM as concerned only with operations and maintenance would be limiting the broad specialisation of the discipline. FM has been viewed in many ways over the years. Atkin and Brooks (2015) described FM as "An integrated approach to operating, maintaining, improving and adapting the building and infrastructure of an organisation to create an environment that strongly supports the primary objectives of that organisation." The concept of operational FM is aimed at organising and the administration processes. This involves activities ranging from controlling, monitoring, and operating, to conform to the policies, plans, and strategies that are in place to ensure that the objectives of the organisation are met.

History and definition of facilities management

Facilities management can be traced back to the Romans who adopted the practice as a necessity (Jawdeh, 2013). The word facility is a medieval French intermediary derivative from the Latin *facio* and *facilis*, which means "I do" and "easy to do", respectively (Bröchner, 2010). During the early Roman Empire and the Late Republic, in about the first century BC till two centuries after, authors from Rome were writing on early ideas of management of facilities (ibid.). These include papyrus letters as well as archaeological evidence that depict what is termed facilities management. There is no proof of the existence of the FN profession at that time; however, there are three sites where coordinated facility services can be identified. These sites spanned across estates involving housing and agricultural production and baths. The management of public baths needed dedicated skills, which led to the appearance of facilities managers. In Roman society. public baths were an important facility and complex to manage, hence the need for facilities managers. Also, Fagan (1999) states that early societies in Portugal engaged in the

DOI: 10.1201/9781003376262-4

management of contracts involving baths; this entailed the services of a facility manager to carry out the washing, drying, and greasing of the bronze implements.

FM has existed for centuries; however, the development of FM as a profession commenced around the 1900s when scientific management, coupled with its deployment in carrying out office administration, emerged (Svensson, 1998). This is corroborated by Wiggins (2014), who states that FM was applied before the 1970s by the US government, the Army, and educational establishments adopting procedures that relevant professionals have now accepted as FM. Though it has long been practised, the term "facilities management" came to prominence in about 1970 (Price, 2003). Since then, the term has become normal for professionals and the like. Becker (1990) observed that the discipline called FM developed because of the following problems:

- an increase in the cost of built space;
- an upsurge in global competitiveness;
- the rise of information technology (IT);
- defects in quality delivery;
- glaring rise in staff expectations.

Also, Solomon and Cloete (2006) define FM as "the coordination of the physical workplace with the needs or objectives of the organisation", while Chotipanich (2004) defines it as "the support function coordinating physical resources and workplace, and support services to user and process of works to support the core business of the organisation". According to Nutt (2004), FM is "the management of infrastructure resources and services to support and sustain the operational strategy of an organisation over time". Equally, FM is taken to be "the process by which an organisation delivers and sustains a quality working environment and delivers quality support services to meet the organisation's objectives at best cost" (Amaratunga and Baldry, 2002).

According to the South Africa Facilities Management (SAFMA) (2012), FM is defined as "a process that enables sustainable enterprise concerning the whole life management of workplaces to achieve productivity and support business effectively". Also, it is described as "a practice that ensures effective operational management of buildings, in both public and private organisations, comprising of a broad range of activities from strategic, operational planning to daily physical maintenance, cleaning and the management of environmental performance issues" (Facilities Management Association of Australia, 2019: 5). The European Facility Management Network (EuroFM) (2019) outlines FM as "the integration of various processes within an organisation to maintain and develop services which support and improve the effectiveness of the organisation's primary activities". Furthermore, the International Facilities Management Association (IFMA) and the British Institute of Facilities Management (BIFM) (2019) include the combination of people in a working environment and other notable processes in their definitions. The definition of FM by both bodies is: "the incorporation of multiple disciplines to ensure the functionality of the built environment by integrating people, process and technology" (IFMA, 2007: 2); and

> the integration of processes within an organisation in the built environment to maintain and develop the agreed services which support and improve the effectiveness of the organisation's primary activities and management of the impact of these processes upon and the workplace.
>
> (BIFM, 2019: 1)

Functions of facilities management

According to Atkin and Brooks (2015), the FM industry has matured and gained a vantage position, offering germane added-value enhancement in pursuing and attaining an organisation's core business. This task can be bolstered by the capabilities of technological innovations and efficient management systems (IFMA, 2007).. FM as a discipline contributes 5 per cent of the world's Gross Domestic Product (GDP) and entails more than just operational procedures but also extends to the establishment and maintenance of a comfortable and productive environment for work processes and activities (Best et al., 2003). Hence, the functions and roles of facility managers significantly affect the economy of any nation through their contribution to GDP, a key economic indicator. This is achieved in two ways: (1) the improvement of the long-term economic value of the nation's infrastructural assets by maintenance processes and policies, caretaking procedures, and adaptation; and (2) improvement of workers' delivery by ensuring health and safety best practices by creating a conducive work environment (Kamarazaly et al., 2013). Also, there is a shift in the focus of FM towards the overall strategic management of facilities, entailing the management of human resources, quality management, and portfolio management of real estate.

The role of facility managers is rapidly evolving into the management of enormous and sophisticated facilities in the built environment. These roles are not limited to the organisation, control, and coordination of the strategic management of facilities and buildings in private and public establishments to ensure efficient and adequate operations during the use of the facilities, coupled with creating a safe environment for users (Booty, 2009). The synergy of these roles brings about the incorporation of a tactical, strategic, and efficient operational system. According to Then (2003), FM can be classified into three types from the administrative perspective: (1) strategic; (2) operational; and (3) tactical FM. Strategic FM dwells on the seamless synergy between facilities and corporate goals. The requirements are the strategic alliance of the setting up and operations of facilities to meet the challenges of the organisation or business concern. To further this, placing FM on a strategic pedestal has a huge influence on the process of decision-making, as it incorporates adopted plans and communication with the top management hierarchy who have the responsibility of taking decisions to ensure that the core business objectives of the organisation are met (Kamarazaly et al., 2013). Alexander (2003) highlights the following as the strategic FM roles:

- creation of facilities policies and its communication;
- planning and implementation of enhanced business services and delivery;
- identification of business requirements and user needs;
- negotiation of business layer agreements;
- putting in place viable contract and contract policies;
- establishment of business partnerships;
- appraisal of the attainment of core business goals.

The FM unit of any organisation has the responsibility of managing the properties/facilities to attain a wide range of objectives, which includes production at an optimal level, continual enhancement of quality, cost reduction, risk abatement and getting value for invested finances (Kamarazaly et al., 2013). Furthermore, it targets the management of corporate assets to provide a conducive environment for the attainment of the organisation's core business goals (Tobi et al., 2013).

Wiggins (2014) states that issues of technical concern affecting the maintenance of facilities and personnel operations cover: keeping up with developments in information technology, maintenance of internal and external components of buildings to aid activities relating to research, management of obsolete building components, and meeting the increasing safety and environmental demands. Also, it is germane that facilities managers ensure the compliance of maintenance management with statutory demands. Operations management entails the activities related to operating facilities and their related services, coupled with the maintenance activities associated with the components of the building or facility. According to the IFMA (2007), within the scope of operations management are administration and periodic maintenance of facilities and associated services, such as the provision of utilities, fire safety and protection, custodial duties, and insurance.

The key issue encountered in operations management is the competence of the facilities manager to handle the repairs or maintenance associated with operations in the facility, coupled with meeting the occupants' or users' demands that might necessitate effective unplanned maintenance and emergency maintenance. Furthermore, Wiggins (2014) noted that space management is based on the setting-up of standards required for space use, the catalogue of space already in use, assessment of the use of space, allocation of space, planning, and supervision of space. Moreover, it is an interdisciplinary endeavour that combines space, technologies, activities, and users in the planning and management of a business or living environment to support the main goals of the business (Li et al., 2017).

Wustemann and Booty (2009) observed that space management is the fundamental display of an organisation's corporate image. The key concern of space management is ensuring that facilities, in conjunction with set-out plans and the use of space, conform to statutory and legal requirements and the building codes related to planning standards, health, and occupancy. Studies of the management of the workplace show that space management can serve as a vital support to the business's core goals and meet the requirement of the users (Blakstad and Torsvoll, 2010). The functions of management are versatile and include the allocation and effective usage of space in a business entity, organisation, or residence. This can be in the form of a single or multiple floors, and the facilities manager must have a better understanding of how best the space of the facility can be put to use. This seeks to answer questions pertaining to what space to use, who is to occupy the space, how much the space will cost, and when the space is to be occupied. According to Atkin and Brooks (2015), to manage space efficiently, the following are recommended:

- optimisation of space based on an outline of a new facility;
- matching developed use to a restored or refurbished facility;
- enhancing usable to gross floor area ratio;
- integrating design attributes to support several functions at various times;
- provision of space, fittings/furniture having adaptability attributes;
- creation of spaces that incorporate open-plan and other types of activities, such as meetings;
- provision of access to wireless data to enable the optimum use of common space.

The guiding principles of space management as a function of FM have been the focus of many studies, as evident in the drive for the actualisation of FM objectives (Wen et al., 2021). However, there are still some prevailing challenges facing effective space management in the delivery of FM functions. Fawcett (2009) notes that the application of mathematical models to calculate the optimal capacity and optimal loading is based on principles related to the uncertainty of demand, displacement cost, and surplus capacity cost. Furthermore, May (2014) noted that immense benefits could be derived from the right application of space management in facilities, ranging from little to no wasted space, energy consumption management, and effective work procedures.

Facility performance

The aim of performance measurement involves identifying both the extent to which a facility meets occupants' needs and the critical factors impacting its performance (Koleoso et al., 2017). Adeyeye et al. (2013) highlight four obligations when assessing performance: (1) criteria for measurement; (2) measurement comparison; (3) measurement assessment; and (4) feedback. The techniques for measurement are initially applied to criteria and to personal scenarios. These measurements aid an objective comparison between varying scenarios. This results in the improvement of communication and helps in decision-making. Developing an innovative course of action to solve problems is done by evaluating scenarios. In completing the cycle, feedback enhances accomplishment by providing ways to improve on previous policies of the decision-makers.

The measurement of facilities' performance mainly entails three elements: (1) the physical; (2) the function; and (3) financial constituents (Lavy et al., 2010). These three elements form the spine for assessment of the performance of facilities. However, Valins and Salter (1996) noted that up to 90 per cent of the effect on the main objectives of an organisation arises from the functional performance of the user facility. The building fabric, including other physical features, is an important physical aspect of the facility (Loosemore and Hsin, 2001). These physical features are energy efficiency, structural integrity, maintainability, heating, lighting, and durability. The functional constituent entails issues involving the building and the occupants, including ambiance, spatial outlay, and health and safety. Assessment of financial performance consists of computing expenditures and depreciation over time. Adeyeye et al. (2013) observed that the recent advances made in post-occupancy evaluation further aid building performance evaluation (BPE) as well as universal design evaluation (UDE), in which factors of a non-technical inclination that impact building designs are valid.

The scope of FM is not confined to the reduction of the cost of operations. However, it should focus on improving building efficiency and equally be adept at assessing the control service required for the successful operations of the business concern (Amaratunga and Baldry, 2002). Lavy et al. (2010) noted that to attain anticipated performance, the measurement of the effectiveness of FM is derived by gaining full knowledge of the prevailing conditions of the facility before the modifications of the FM services. Also, Then (2003) stated that the measurement of facilities' performance is an emerging area within strategic FM and forms a nexus between FM and corporate strategies. Thus, in tackling business success and building effectiveness, project eventualities ought to be assessed through the lens of FM. The processes will help in the identification of FM requirements and concerns.

A completed building facility is expected to perform the predetermined functions as outlined at the conceptualisation stage of the project. Stanley (2001) noted that owners of buildings and likewise its users look for the optimal delivery of completed buildings to support the organisation's mission, to enhance the productivity of workers, to improve profits, promote a good image, ensure comfort, functionality, and safety. On completion, buildings are expected to meet their outlined functions (Idrus et al., 2009). Ideally, buildings are expected to have a long service life (Wen et al., 2021). The eventual delivery performance of buildings is hugely determined by their users throughout their life span (Olanrewaju et al., 2011). The requirements expected to be met by buildings are grouped into statutory, functional, user and performance requirements (Watt, 2007). These requirements can further be streamlined into two broad categories: user and performance requirements:

- *User requirements* refer to suitability for purpose and defence against the external environment, allocation of space, and comfort of the users. To make a building functional, it must meet the set requirements.
- *Performance requirements* refer to statutory requirements related to thermal comfort, appearance, sanitation, durability, security, strength and stability, cost, access, lighting and ventilation, acoustic control, and fire protection.

According to Stanley (2001: 22), building performance can be defined as "the degree to which a building or other facility serves its users and fulfills the purpose for which it was built or acquired; the ability of a facility to provide the shelter and service for which it is intended". Performance of buildings can be viewed from the perspective of expectations to be met by buildings (Haupt, 2001). Furthermore, Williams (1993) stated that this refers to buildings' contributions to meeting the functional requirements and anticipated demands of the occupants over time. In essence, it connotes users' expectations and satisfaction, and requirements (Olanrewaju et al., 2011).

The Australian Department of Treasury and Finance (2005) outlined that it is important that as buildings are being put to use, they should be consistently appraised, based on business strategies and service outputs. Post-occupancy evaluation is a strategic examination of built facilities to ascertain how the facilities are meeting the users' requirements (Fianchini, 2007). The discussion of building performance is becoming prominent due to its influence on the effective use of buildings (Douglas, 1996). Haupt (2001) stated that the assessment of building performance is carried out to make certain that occupants comfortably and permanently carry out their activities safely and the building fulfils the requirements for comfort without adverse effects on users' health.

According to the Department of Treasury and Finance (2005), the objectives of building performance assessment are the following:

- identification of buildings that are underperforming;
- identification of targeted under-performing building elements;
- provision of data to assist in predicting the future performance course;
- determination of suitable standards of maintenance needed for the building.

A four-pronged framework was conceptualized by the Department of Treasury and Finance (ibid.) to assess the performance of buildings. This framework comprises the following constructs: service performance, strategic performance, technical performance, and financial performance.

Assessment of facilities' performance

There is a need for a periodic assessment of the performance of the operations of facilities management concerning its contribution to the core business mandate of the organisation. A variety of management tools are used to assess performance, including the balanced scorecard, performance measurement, benchmarking, and activity measurement (Ashworth and Tucker, 2017). Some of these tools are best suited for the measurement of the performance of FM, while others are adopted for the assessment of customers' satisfaction (Perera et al., 2016). The complementary units have the responsibility of carrying out the core duties of the organisation, and senior management make up the customers of FM. A pragmatic assessment of the performance of FM is used in the measurement of customers' satisfaction. Usually, the delivery outcome of the FM activities serves as the input garnered by other sectors of the organisation, which ultimately serves as a delivery outcome in meeting the business undertakings of the organisation.

Carder (1997) highlighted the "knowledge-base tool of management used in organisations" interface and called it an "informed interface". The management tool requires

> taking the tasks of analyst, adviser, and educator of the customer; this interface role is increasingly needed between the customer and operational management and delivery services. The interface role will be required to understand and use both business and facilities information, combined to create organisation-specific workplace knowledge.

Another distinguishing element of this measurement tool is measuring effectiveness rather than internal competence by the FM process (Hassanain and Iftikhar, 2015). The sophistication of the technology adopted, or the detailing of the organisation's structure, is not particularly pertinent to the senior management but rather it is the efficient delivery of operations and strategic management of systems that yield tangible output (Ashworth and Tucker, 2017). The joint monitoring of the important delivery indicators in conjunction with optimum records management and a review of operations guarantees that:

> [The] FM organisation which creates and continuously updates this new performance knowledge will be equipped to provide the role of an analyst, adviser, and educator … which is increasingly being demanded by the customers. Moreover, FM with this ability will be able to defend their position as operational managers.
>
> (Perera et al., 2016)

According to Perera *et al.* (ibid.), the generic form of representing the contributions of FM in a classic workplace interface is "location, buildings and plant, information technology or transport, people and others". In a typical workplace, the functions of the FM unit can be annotated as "input", whose effective functions ensure the delivery of mandated business operations annotated as "output". Another kind of measurement tools of FM performance is the "balanced scorecard"; this takes the form of an airplane cockpit – it gives a view of complex information for managers at a glance (Hassanain and Iftikhar, 2015). Lavy et al. (2010) highlighted different practices used in evaluating facility performance. In a review of tools for measuring post-occupancy assessment, the following were analysed:

- building quality assessment (BQA)
- serviceability tools and methods (STM)
- post-occupancy review of building engineering (PORBE)
- building research establishment environmental assessment method (BREEAM).

Both BQA and STM are used to assess the requirements for occupants and grade the quality of the building. Meanwhile, PORBE serves as a tool for gathering feedback from occupants of a building facility, and BREEAM is used for assessing the environmental conditions of constructed facilities.

Post-occupancy evaluation (POE)

POE serves as a viable tool adopted for the measurement of building performance as well as occupant satisfaction. RIBA (1991) defines post-occupancy evaluation as "a systematic study of buildings in use to provide architects with information about the performance of their designs and building owners, and users with guidelines to achieve the best out of what they already have". Similarly, the Federal Facilities Council defines it as follows: "post-occupancy evaluation (POE) is a process of systematically evaluating the performance of buildings after they have been built and occupied for some time" (Preiser, 2002: 18). There is a difference between POE and other building performance evaluations as the former emphasises the needs of occupants of the building coupled with aesthetic function, psychological comfort, health and safety, security efficiency, and satisfaction.

POE, sometimes referred to as post-occupancy assessment, is a concept that dwells on a group of activities to collect the necessary information regarding building performance upon its completion and being ready for operations (Hewitt et al., 2005). These activities equally address questions about building occupants and the level of satisfaction related to the immediate environment of the newly built structure. This fully agrees with the definition of POE that it is not centred only on the technical sections or the performance of buildings, but equally focuses on the users' and occupants' needs, and the management of the facility.

According to Hassanain and Iftikhar (2015), specialists in FM know that POEs coupled with their eventualities in different types of buildings play a significant role in the briefing phase of the construction process. This also brings together the varying parties involved, such as clients, designers, and occupants. The goal of achieving occupants' satisfaction and a high quality product can be achieved by the evaluation of buildings and performance measurement, coupled with the output information to make plans for the future. Cohen et al. (2001)

outlined the questions usually put forward in carrying out a POE. These centre on issues to do with the conditions of the physical environment, ranging from thermal comfort, lighting, air quality, and noise control; personal influence over the conditions of the physical environment, response management of complaints, general comfort, and the general building quality.

Benchmarking

Benchmarking evolved from the process of total quality management to aid managers in the application of contextual circumstances about performance evaluation (Dodd et al., 2022). Usually, benchmarking is used to identify a reference point or benchmark, acting as the yardstick for making decisions on performance. The benchmark is either an internal circumstance connected to the organisation, or an external circumstance connected to best practices or the competition. Adewunmi et al. (2016) stated that the aims of benchmarking are to assist the organisation in achieving a clear path on external references, coupled with assisting in identifying the best practices in the industry, when comparisons of other performances are made. Benchmarking is seen to be vital in the assessment of building performance because it includes cost-related and non-cost-related benchmarks. Issues involving the former are quantitative parameters offering short-term opinions, while those concerning the latter are qualitative parameters offering long-term opinions. Table 2.1 presents the process involved in benchmarking building facilities by Leake and Stanley (1994).

Table 2.1 The benchmarking process

Phase	*Procedures*
Planning	Identification of performance parameters of the facility to undergo benchmarking
	Identification of best practices to make comparisons with
	Collection of data on performance.
Examination	Comparison made with best practice and identification of gaps in performance
	Adoption of measures for improving performance
Integration	Communication of results and acquiring approval for the need for improvement
	Setting out goals for performance
	Initiation of plans for action and strategies for implementation
Action	Implementation of decisive actions and progress monitoring
	Re-assessment of benchmark parameters and updating changes

Key performance indicators (KPIs)

A performance indicator is defined by Fitz-Gibbon (1990) as the measurement of performance. Key performance indicators (KPIs) are broad pointers of performance dwelling on important elements of output (Chan and Chan, 2004). Similarly, Meng and Minogue (2011) observed that KPIs had gained ground in various industries, being deployed as a system for performance assessment. Usually, in the construction industry, the principal indicators are cost, time, and quality. The use of KPIs in FM has many advantages (Loosemore and Hsin, 2001). It can indicate the direction of managerial efforts in the delivery of overall performance and be integrated with the services for FM, to explicitly outline the necessary outcomes, coupled with important monitoring and control. Usually, the vast majority of KPIs produced in line with the processes of FM concern operations and maintenance cost, environmental matters, management of space, revenue issues, and safety issues.

Lavy et al. (2010) stated that the presentation of KPIs should be portrayed to signify the performance of facilities in general. KPIs should also be monitored within a given time frame and thereafter make comparisons to a baseline to ascertain changes concerning deterioration or improvement (Cable and Davis, 2004). However, Brackertz (2006) affirmed that emphasis on performance measurement should be placed on business goals and job satisfaction instead of dwelling only on finances, as has been the case in the past. Aside from dwelling on the business and employees, the measurement of performance should depend on facility users, including customers, managers, and supervisors (Dasandara et al., 2022). Varying measures are adopted for different users depending on the target projected by the user type. Yuan et al. (2009) noted that KPIs broadly cover five different topics: (1) a project's physical characteristics; (2) marketing and finance; (3) stakeholders; (4) learning and innovation; and (5) processes. However, Amaratunga and Baldry (2002) categorised it into four broad areas: (1) internal processes of FM; (2) relations with customers; (3) financial implications, and (4) growth and learning. Also, Augenbroe and Park (2005) outlined a different perspective, and posited four categories: (1) lighting; (2), energy; (3) thermal comfort; and (4) maintenance. Furthermore, eight varying categories were highlighted by Hinks and McNay (2005): (1) environment; (2) business benefits; (3) space; (4) change; (5) equipment; (6) consultancy; (7) services/maintenance; and (8) general.

Summary

The importance of FM to any organisation cannot be over-emphasised as it serves as a strategic vehicle for meeting the organisation's stated objectives. This chapter focused on the theoretical perspectives of FM by engaging in a holistic review of the extant literature. The history and different definitions of

FM from various sources were outlined. Furthermore, the functions of FM were presented using theoretical lenses. Also, in assessing facilities' performance, the various yardsticks for ascertaining the performance include post-occupancy evaluation (POE), benchmarking, and key performance indicators (KPIs).

References

Adewunmi, Y., Iyagba, R., and Omiron, M. (2016). Multi-sector framework for benchmarking facilities management. *Benchmarking: An International Journal*, 24(4), 109–124.

Adeyeye, K., Piroozfar, P., Rosekind, M., Winstanley, G., and Pegg, I. (2013). The impact of design decisions on post-occupancy processes in school buildings. *Facilities*, 31(5/6), 255–278.

Alexander, K. (2003). A strategy for facilities management. *Facilities*, 21(11), 269–274.

Amaratunga, D. and Baldry, D. (2002). Performance measurement in facilities management and its relationships with management theory and motivation. *Facilities*, 20(10), 327–336.

Ashworth, S. and Tucker, M. (2017). *FM Awareness of Building Information Modelling (BIM)*. 1st edn. BIFM.

Atkin, B. and Brooks, A. (2015). *Total Facility Management*. 4th edn. Oxford: Wiley-Blackwell.

Augenbroe, G. and Park, C.S. (2005). Quantification methods of technical building performance. *Building Research and Information*, 33(2), 159–172.

Becker, F. (1990). *The Total Workplace: Facility Management and Elastic Organisation*. New York: Van Nostrand Reinhold.

Best, R., Langston, C., and De Valence, G. (2003). *Workplace Strategies and Facilities Management: Building Value*. Oxford: Butterworth-Heinemann.

Blakstad, S.H. and Torsvoll, M. (2010). *Tools for improvements in workplace management*. In Proceedings of the 9th EuroFM Research Symposium, Madrid, Spain, pp. 1–14.

Booty, F. (2009). *Facilities Management Handbook*. 4th edn. Oxford: Butterworth-Heinemann.

Brackertz, N. (2006). Relating physical and service performance in local government community facilities. *Facilities*, 24(7/8), 280–291.

British Institute of Facilities Management (BIFM) (2019). About facilities management. Available at: http://www.bifm.org.uk/bifm/about/facilities (accessed 30 October 2020).

Bröchner, J. (2010) Innovation and ancient Roman facilities management. *Journal of Facilities Management*, 8(4), 246–255.

Cable, J.H. and Davis, J.S. (2004). Key performance indicators for federal facilities portfolios, Federal Facilities Council Technical Report No. 147. Washington, DC: National Academies Press.

Carder, P. (1997). The interface manager's toolkit. *Facilities*, 15(3/4), 84–89.

Chan, A.P.C. and Chan, A.P.L. (2004). Key performance indicators for measuring construction success. *Benchmarking: An International Journal*, 11(2), 203–221.

Chotipanich, S. (2004). Positioning facility management. *Facilities*, 22(13), 364–372.

Cohen, R., Standeven, M., Bordass, B., and Leaman, A. (2001). Assessing building performance in use 1: The Probe process. *Building Research and Information*, 29(2), 85–102.

Collings, T. (2007). *Building and operations management*, Paper presented at the NZPI CPD Seminar, Wellington, NZ, April.

Dasandara, M., Dissanayake, P., and Fernando, D. (2022). Key performance indicators for measuring performance of facilities management services in hotel buildings: A study from Sri Lanka. *Facilities*, 40(5/6), 316–332.

Department of Treasury and Finance (Government of Western Australia). (2005). Maintenance policy. Available at: http://www.treasury.wa.gov.au/cms/uploadedFiles/04_samf_mp_082005.pdf (accessed 26 October 2020).

Dodd, J., Smithwick, J., Call, S., and Kasana, D. (2022). The current state of benchmarking use and networks in facilities management. *Benchmarking: An International Journal*, [Ahead of Print].

Douglas, J. (1996). Building performance and its relevance to facility management. *Facilities*, 14(3/4), 23–32.

European Facility Management Network (EuroFM) (2019). What is facility management? Available at: http://www.eurofm.org/index.php/what-is-fm?showall=&start=2 (accessed 17 October 2020).

Facilities Management Association of Australia (FMAA) (2019). Available at: http://www.fma.com.au/cms/index.php?option=com_content&task=view&id=45&Itemid=5 (accessed 30 October 2020).

Fagan, G.G. (1999). *Bathing in Public in the Roman World*. Ann Arbor, MI: University of Michigan Press.

Fawcett, W.H. (2009). Optimum capacity of shared accommodation: Yield management analysis. *Facilities*, 27(9/10), 339–356.

Fianchini, M. (2007). A performance evaluation methodology for the management of university buildings. *Facilities*, 25(3/4), 137–146.

Fitz-Gibbon, C.T. (1990). *Performance Indicators*. Clevedon, PA: Multilingual Matters.

Hassanain, M. and Iftikhar, A. (2015). Framework model for post-occupancy evaluation of school facilities. *Structural Survey*, 33(4/5), 322–336.

Haupt, T.C. (2001). The construction approach to construction worker safety and health, PhD thesis, University of Florida, Florida.

Hewitt, D., Higgins, C., and Heatherly, P. (2005). A market-friendly post-occupancy evaluation: building performance report: Final Report – Contract C 377 10091. New Buildings Institute, Inc., prepared for: Northwest Energy Efficiency Alliance, Portland, OR.

Hinks, J. and McNay, P. (1999). The creation of a management-by-variance tool for facilities management performance assessment. *Facilities*, 17(1/2), 31–53.

Idrus, A., Khmidi, M., and Abdul Lateef, O. (2009) . Value-Based Maintenance Model for University Buildings in Malaysia: A critical review. *Journal of Sustainable Development*, 2(3), 127–133.

International Facilities Management Association (IFMA) (2007). Exploring the current trends and future outlook for facility management professionals. Available at: http://www.ifma.org/what_isfm/index.cfm (accessed 13 October 2020).

Jawdeh, H. (2013). Improving the integration of building design and facilities management. Unpublished PhD thesis, University of Salford, UK.

Kamarazaly, M.A., Mbachu, J.I., and Phipps, R. (2013). Challenges faced by facilities managers in the Australian universities. *Journal of Facilities Management*, 11(2), 136–151.

Koleoso, H., Omirin, M., and Adewunmi, Y. (2017). Performance measurement scale for facilities management service in Lagos-Nigeria. *Journal of Facilities Management*, 15(2), 128–152.

Lavy, S., Garcia, J.A., and Dixit, M.K. (2010). Establishment of KPIs for facility performance measurement: Review of literature. *Facilities*, 28(9), 440–464.

Leake, E. and Stanley, J. (1994). *Benchmarking for Facility Management Workbook*. International Facility Management Association – Benchmarking Management.

Li, L., Yuan, J., Ning, Y., Shao, Q., and Zhang, J. (2017). Exploring space management goals in institutional care facilities in China. *Journal of Healthcare Engineering*, 6307976, 1–16.

Loosemore, M. and Hsin, Y.Y. (2001) . Customer-focused benchmarking for facilities management. *Facilities*, 19(13), 464–476.

May, M. (2014). *Modeling and Optimisation in Strategic Space Management*. FMJ IFMA.

Meng, X. and Minogue, M. (2011). Performance measurement models in facility management: a comparative study. *Facilities*, 29(11/12), 472–484.

Nutt, B. (2004). New alignments in FM. *Facilities*, 22(13), 330–334.

Olanrewaju, A.L., Khamidi, M.F., and Arazi, I. (2011). Appraisal of the maintenance management practices of Malaysian universities. *Journal of Building Appraisal*, 6(3/4), 261–271.

Perera, B., Ahamed, M., Rameezdeen, R., Chileshe, N., and Hosseini, M. (2016). Provision of facilities management services in Sri Lankan commercial organisations – Is in house involvement necessary? *Facilities*, 34(7/8), 394–412.

Preiser, W.F.E. (2002). Continuous quality improvement through post-occupancy evaluation feedback. *Journal of Corporate Real Estate*, 5(1), 42–56.

Price, I. (2003). Facility management as an emergent discipline. In R. Best, C. Langston, and G. de Valence (eds), *Workplace Strategies and Facility Management: Building in Value*. Oxford: Butterworth-Heinemann.

RIBA (1991). *Architectural Knowledge: The Idea of a Profession*. London: E. & F.N. Spon.

Solomon, L.A. and Cloete, C.E. (2006). *The effectiveness of facilities management services in the Western Cape*. In Proceedings of the 1st Built Environment Conference, Association of Schools of Construction of Southern Africa, pp. 306–313.

South African Facilities Management Association (SAFMA) (2012). What is FM? Available at: http://www.safma.co.za/portals/0/What_is_FM_presentation.pdf (accessed 28 October 2020).

Stanley, L. (2001). *Federal Facilities Council Technical Report No.145: Learning from Our Buildings: A State-of-the-Practice Summary of Post-Occupancy Evaluation*. Washington, DC: National Academy Press.

Svensson, K. (1998). Integrating facilities management information a process and product model approach. Thesis. The Royal Institute of Technology. Available at: http://citeseerx.ist.psu.edu/viewdoc/download?doi=10.1.1.198.139&rep=rep1&type=pdf (accessed 16 September 2021).

Teicholz, E. (2001). *Facility Design and Management Handbook*. New York:McGraw-Hill.

Then, D.S.S. (2003). Strategic management. *In* R. Best, C. Langston *, and* G. De Valence *, (eds), Workplace Strategies and Facilities Management: Building In Value*, Oxford: Butterworth-Heinemann, pp. 69–80.

Tobi, S., Amaratunga, D., and Noor, N. (2013). Social enterprise applications in an urban facilities management setting. *Facilities*, 31(5/6), 238–254.

Valins, M.S. and Salter, D. (1996) . *Futurecare: New Directions in Planning Health and Care Environments*. Oxford: Blackwell Science.

Watt, D.S. (2007). *Building Pathology*. 2nd edn. Oxford: Blackwell Publishing.

Wen, Y., Tang, L., and Ho, D. (2021). A BIM-based space-oriented solution for hospital facilities management. *Facilities*, 39(11/12), 689–702.

Wiggins, J. (2014). *Facilities Manager's Desk Reference*. 2nd edn. Chichester: John Wiley & Sons.

Williams, B. (1993). What a performance! (Editorial). *Property Management*, 11(3), 190–191.

Wustemann, L. and Booty, F. (2009). Space design and management In F. Booty (ed.), *Facilities Management Handbook*, 4th edn. Oxford: Butterworth-Heinemann.

Yuan, J., Zeng, A.Y., Skibniewski, M.J., and Li, Q. (2009). Selection of performance objectives and key performance indicators in public-private partnership projects to achieve value for money. *Construction Management and Economics*, 27(3), 253–270.

3 Digital technologies for facilities management

Introduction

The use of digital technologies in almost all spheres of human endeavour is fast gaining recognition. This is spurred by the ability of these digital technologies to enhance productivity, increase efficiency, aid in the timely completion of tasks, enhance sustainability, and improve competitive advantage, among others. In architecture, engineering, construction, and facilities management (FM), much has been said about the slow adoption of innovative technologies. This delay has impeded the optimisation of the processes in delivering the mandates of these sectors. For the construction industry, in particular, the tag of "slow adopter" of innovations is constantly being attributed to it. Likewise, FM activities have been perennially associated with conventional or traditional modes and methods of delivery. However, with the advent of the Fourth Industrial Revolution, there is a gradual shift towards integrating digital technologies into FM processes. In this chapter, a review of some of the digital technologies used by FM is presented. A detailed review of how these digital technologies can be applied in the effective delivery of FM is offered, while also unravelling the potential benefits to be accrued from these digital technologies. Also, the challenges posed by implementing these digital technologies are explored with a view to suggesting mitigating measures to these challenges.

Digital technologies to benefit FM

Computer-aided facilities management (CAFM)

Generally, the term "computer-aided facilities management" is used to describe all kinds of information systems used in FM. This system aims to connect database technology with graphic-aided design systems to facilitate the seamless management of built facilities. Basic application coverage includes computerised maintenance, room scheduling, asset management, lease tracking, space management, among others.

The choice of the right tool ought to be based on the aspirations and goals of the organisation (Rycroft, 2006). These systems need to be seen by the

DOI: 10.1201/9781003376262-5

facilities manager as tools (Araszkiewicz, 2017), the selection and operation of which should be carried out by considering the optimum results and delivery (Aziz et al., 2016). Technological tools for FM are rapidly becoming popular, thus enabling organisations to leverage the values of FM, such as the creation of FM strategic plans, managing property portfolios, responding to service requests, verifying data, and communicating with other organisations. Aziz *et al.* (ibid.) noted that the adoption of information technology for FM purposes enables organisations to occupy a vantage position in core business undertakings by effecting change innovation and management, coupled with improving efficiency and offering the likelihood of new services for users.

According to Araszkiewicz (2017), the activities covered by software for FM and operations include inventory control, asset record creation, and maintenance, work order formulation, plan scheduling and development, report creation, and invoice match-up. Such software enables a whole operations and maintenance schedule that aids in the inconspicuous attainment of predetermined FM goals. This is supported by information model tools that create a platform for importing generated data to a central computerised maintenance management system (CMMS), which is enabled by graphical spatial information (Akcamete et al., 2010). Although Gnanarednam and Jayasema (2013) stated that computer-aided facilities management (CAFM) is largely dependent on conventional methods, including tabulated data and 2D graphics. But it still goes a long way in the management and organisation of facilities, such as facility budgeting, operations and management, space inventory, project management, interior planning, lease management, and equipment management (Lee et al., 2013).

Building Information Modelling (BIM)

Recent technological innovations have led to the integration of Building Information Modelling (BIM) in the process of FM. Volk et al. (2014) state that the introduction of BIM was facilitated in the early 2000s to aid architects and engineers steer project delivery. However, the introduction of BIM into FM processes commenced in about April 2005 in Sydney, Australia (CRC Construction Innovation, 2007). The emergence of BIM technology in FM is centred on more information management regarding the building structure and human capital. Sabol (2008) noted that BIM functions are based on the simulation of the components of a building and portray virtual reality. Due to the demands of stakeholders, BIM aids the entire process involved in construction, spanning design, actual construction, maintenance, and demolition, all dependent on the level of development (LOD) (Volk et al., 2014). The development of the BIM model has six levels (e.g. LOD 350 and LOD 500) in the determination of model elements to be included. This helps the project stakeholders specify the information with a high level of reliability and clarity for the model (BIM Forum, 2015). The development of a BIM to LOD 500 ought to be considered for FM, as it is characterised by geometry and information to aid operations and maintenance activities, such as production capacity, inventory, workflow, and safety (Love et al., 2013). The use of BIM for FM functions supports the management of life-cycle data and performance monitoring of activities involved in FM, such as managing emergencies, energy performance, quality control, space management, maintenance and service, or the development of a plan for preventive maintenance (Becerik-Gerher et al., 2012).

There are various benefits derived from the adoption of BIM in FM, such as the provision of "as-built" information, which is valuable (Eastman et al., 2011), maintenance records, service details (Chen et al., 2011), quality measurement (Boukamp and Akinci, 2007), assessment and monitoring (Abdullah et al., 2014), emergency procedures (Arayici, 2008), reduced error rate, timely project delivery, reduction of costs, and providing feedback for the elimination of design-related performance issues (Oti et al., 2016), and retrofit planning (Arayici, 2008). This allows facility managers to access important information from a single source of electronic files, thus eliminating the cumbersome process of digging for information from a bulky body of data (Azhar et al., 2012). Furthermore, BIM gives object-based 3D visualisation data, which is pertinent in managing FM data. Also, the acquisition and identification of data are more efficient and faster when deployed for FM systems that are object-based in collaboration with BIM (Kim and Hong, 2017). Other benefits of the fusion of BIM with FM practices include improvement in the efficiency and productivity of employees, better response time during operations, improvement in interoperability, automatic model updates, an easier process to retrieve data, requirements combined with contract documents, provision of better FM requirement information for design and construction, seamless attainment of the integration of life-cycle, provision of a holistic strategic plan, and collaboration improvement (Terreno et al., 2015).

Gu et al. (2008) highlighted one particular benefit of integrating BIM into FM is the improvement in handover. The documentation of most contracts involves the compilation of a great deal of information, such as a list of equipment, preventive maintenance schedules, warranties, a list of spare parts, sheets of product data, etc. This set of information is vital to owners and property managers since it serves as a basis for the support of FM activities; however, the handover of information is carried out manually in the FM phase. This poses many difficulties as facilities managers are usually hampered in accessing and using some of the information given during the handover process as some pieces of data can be inaccurate and incomplete (Kelly et al., 2013). According to East (2013), the Construction-Operation Building Information Exchange (COBie) stipulates the standards and data required for operations and maintenance (O&M) and project handover. The building data on layers, systems, structures, and spaces can be digitised and be stored in BIM models in the entire life-cycle of the building. This reduces time and costs to re-enter data for FM, plus the improvement of quality, efficiency, and reliability of the data (Teicholz, 2013).

Furthermore, BIM can make predictions on the performance of buildings and control life-cycle costs via budgeting based on the collected data of the building and the environment. BIM also can provide specific data for various users and several needs. The use of BIM can help in the analysis of building proposals and engage in the simulation and benchmarking of the performance of buildings, coupled with the facilitation of the parametric building analysis and semantic relationships (Atkin and Brooks, 2009). Informed decisions can be made regarding the improvement in the delivery of facilities that FM practitioners can make by extracting and analysing the data (Azhar et al., 2012). Also, Becerik-Gerher et al. (2012) noted that detecting broken equipment, system faults, and lack of comfort conditions and adjusting ergonomics are seamlessly achieved when BIM is integrated into FM functions. Moreover, other important applications are energy monitoring and control, safety monitoring, space management, and staff training (Kelly et al., 2013). Due to the complexities of building designs, defects arising from designs are hard to visualise, leading to enormous challenges when engaging in maintenance activities. The combination of BIM, FM specifications, and facility managers' knowledge can be incorporated into constructing projects (Liu and Issa, 2016). This combines design analysis and simulation of performance for project upgrades and refurbishment (Sabol, 2008). The building operations provided by BIM can serve as a basis for the provision of future designs (Building Information Modelling (BIM) Task Group, 2012).

Systems to support FM, such as CAFM, can be integrated with BIM. CAFM has the capabilities to incorporate full asset information of facilities, which saves time in the input of data; also, the accuracy and quality of the data can be confirmed, thus leading to reduced running costs and prompt attainment of optimum performance (ibid.). Workers are prone to accidents and injury when engaged in maintenance and repair work in FM. Hence, the task of FM demands safety data coupled with the necessary safety procedures for its application. This can easily be accessed and retrieved from BIM integrated into the functions of FM at the post-occupancy stage of a building's life-cycle.

Internet of Things (IoT)

The Internet of Things (IoT) application is fast being adopted as it permits the rapid collection, transmission, and exchange of data through embedded sensors and wireless technologies (Sidek et al., 2022). Radio Frequency Identification (RFID) is considered an "enabling technology" in the IoT. It entails four rudimentary constituents: tag, reader, software, and computer network (Affia and Aamer, 2022). For FM activities, sensor networks and RFID have immense potential for use in functions. These include building life-cycle constituents tracking, control of building energy, inventory management, document management, building security, provision of efficient and cost-effective logistics, and tracking of materials (Gubbi et al., 2013). The recognition of facilities items requiring repair or maintenance through location detection of RFID has great potential to improve the facilities maintenance management (Becerik-Gerber et al., 2012). Ko (2017) formulated an FM system characterised by a web-based RFID that permits the automatic recognition of facility identity. The system is connected to the internet, and permits data management, forecasting, scheduling, and statistical analysis. The engagement of maintenance tasks by users is permitted through a web-based system. Building maintenance activities are carried out in work orders which can be arranged spontaneously through scheduled modules, thereby aiding in the reduction of operations duration, avoiding duplicate or misplaced maintenance activities.

Also, the system has a fuzzy neural network in its forecast module that permits the forecast of building components' lifetimes, minimising the possibility of breakdowns and malfunction. The accuracy in detecting the pattern of building occupancy is considered vital in attaining the right decisions for building energy control and service delivery (Newsham et al., 2016). Sensor devices are installed around the building to monitor the consumption behaviour of the building and control or detect anomalies (Araya et al., 2017). The monitoring and visualisation of the records and status of facilities can be carried out through the system's statistical module. However, many existing buildings are not designed for or incorporate sensory systems, thus hindering occupancy detection, and measurement and monitoring of building performance.

Recent IoT technologies give building systems the ability to undergo modification and upgrades by adding capabilities involving implicit occupancy sensing. Melfi et al. (2011) classified the implicit occupancy sensors into three tiers (Tiers I, II and III), based on the level of modification work required, using the existing data on occupancy. The advent of the IoT, emphasising RFID technology, is an innovative platform for building facility analytics and data storage. Lin et al. (2014) outlined a platform for sharing maintenance information by applying an RFID system to maintenance management and a portable 2D barcode to improve a lab's facility maintenance management. Furthermore, Motamedi and Hammad (2013) assessed the capabilities and localisation of several hand-held RFID technology networks. To identify the location of movable assets within the facility and fixed assets on the floor plans, the neighbourhood method was used. Progress in the application of RFID for FM has brought about the fusion of the BIM database with RFID, which captures activities and information on building maintenance and provides better building monitoring and performance. However, despite these laudable innovations, a major challenge is the noise from the logged data which needs to be tackled from the perspective of the current RFID. A typical example is a study carried out by Ergen et al. (2007), which noted that the accuracy of indoor object detection by RFID in facilities was influenced by various obstructions, including metallic objects. Compared to outdoor construction site environments, indoor environments have more complexity and uncertain extracted scenarios, thereby making them prone to mistakes during the process of information extraction.

Geographic Information Systems (GIS)

There is a growing acceptance of the application of Geographic Information Systems (GIS) technology for FM. The application of GIS in FM is centred on managing the external works of construction projects and airports, such as electricity supply, municipal water and wastewater (Meerow et al., 2016). The geospatial data from the GIS aids the management of facilities to support various information systems and business processes. A good example is an analysis carried out by facility managers on space availability, utilisation and optimisation over a given area. The data retrieved aids various types of inspection and assessments in the building. Also, the fusion of GIS-based tools and BIM has resulted in innovation in GIS application in the management of facilities. Wu et al. (2014) carried out a combination of BIM, GIS and cloud-based systems technology to facilitate building energy performance through real-time detection. The capacity of building information that is locally based with wireless sensor networks is improved by the system. Karan and Irizarry (2014) formulated a GIS-BIM combined system that incorporates models in 3D semantics with the supply chain involved in the management of facilities. Moreover, Kang and Hong (2015) established a software architecture based on GIS-BIM integration for the efficient extraction, conversion, and loading of information of facilities. Also, the use of IFC geometrical information presented a streamlined basic model that permits GIS-based objects to be visualised. Kang et al. (2016) outlined a combination of GIS-based data and a BIM framework (BG-DI), the integration of which is targeted at enhancing diverse information interpolation from GIS, BIM, and systems of FM.

Virtual reality

Over time, the advances in FM processes have been tasked with attaining cost and time efficiencies, effective methods for surveying infrastructure and building stock, and the automated generation of as-built models. Reality capture technology is one of the fast-growing digital technologies that are propelling the generation of reliable information on the condition of a facility. The models generated comprise the spatial data from the building formulated in data sets based on a series of 3D Cartesian point cloud data (Amano and Lou, 2016). The processing and transfer of the collected points are performed in a structure or purposely designed, although there are still challenges facing the use of cloud technology in the management of facilities. This includes the geometrical coordination of point cloud data, entailing intensity values without semantic data (ibid.). Also, Fathi and Brilakis (2011) established a method for generating Euclidean 3D point cloud data within a campus building, based on automated stereo vision, using video streams captured with a camera characterised with two calibrations. Furthermore, Klein et al. (2012) assessed the processing of photogrammetric image applications to documents and also in verifying the as-built parameters of completed buildings. However, there is still work to be done on the improvement of surveys that are image-based in the process of digital building documentation automation. Hichri et al. (2013) established a method that enriches 3D models of historical buildings, usually the conceptual phase of the collection of data and segmentation, thereby bringing about a final BIM that is well structured and can efficiently engage in modification, conservation, restoration, or maintenance of projects.

Moreover, Xiong et al. (2013) outlined a method for automating the conversion of 3D point data into a compact semantic model. This approach poses challenges, such as surface model conversion to volumetric models. In an attempt to extend cloud point data application to a more robust indoor environment, Jung et al. (2014) established an approach with semi-automation capabilities using models of as-built BIM. The method leads to a reduction in point cloud data size without information loss. This approach helps create as-built BIMs for structures that are characterised as complex and huge; however, one particular problem is the noise and quality control in the cloud's point. Automation or productivity processes in the creation of as-built BIM still need to be improved.

Challenges to the digitalisation of FM

A lot of challenges still face the effective delivery of FM functions. Although there have been recent technological advances to help mitigate most of these challenges, as shown in the preceding section. However, even these digitalised functions in FM delivery are not immune to some of the problems facing the effective delivery of FM activities. According to Kassem et al. (2015), for BIM, the creation and use of as-built models are found in a handful of FM projects, but updating the models to reflect changes in the physical facility is rarely done. The consequences are the production of an inaccurate model of the facility and the loss of the savings in cost and time connected with the ability to locate the elements of the building through the model. Also, Becerik-Gerber et al. (2012) observed that this might lead to organisational issues whereby the responsibilities for model maintenance will not be correctly allocated during handover. In ensuring that the model continues to provide meaningful and usable data, the responsibilities must be designated quite early to prevent redundancy of the model over time (Kassem et al., 2015).

Furthermore, cultural barriers might be an impediment to the use of digital technologies for FM. The construction and FM industries have been criticised as unreceptive to new technologies, which means that the inherent benefits of the uptake of such innovative technologies is hampered. Also, the combined lack of skills and awareness of these innovative technologies hinders their utilisation (Shen et al., 2016). Since most facilities managers are not equipped with the requisite skills to use innovative technologies, it is natural not to use these technologies in the delivery of FM functions. Similarly, Asare et al. (2021) stated that legal barriers and the challenge of model ownership are also impediments to infusing digital technologies into FM, while Kassem et al. (2015) noted that interoperability concerning standardisation of various innovative technologies is challenging to FM functions. The lack of clear requirements in implementing these innovative technologies is also a challenge to their use (Shen et al., 2016). To overcome this challenge, Succar et al. (2016) proposed using deliverables that are model-based to clarify the outlined requirements.

Moreover, Kassem et al. (2015) noted that the lack of methodologies to outline the inherent benefits accompanying the use of these digital technologies is also a factor restricting their espousal. At the same time, return on investment (ROI) appears to be a stumbling block for digitalising FM activities. Shen et al. (2016) stated that due to the fact that beneficial gains resulting from adopting digital technologies for FM functions take a long time, there is a reluctance on the part of stakeholders to adopt these technologies. This challenge, coupled with the significant intangible benefits accompanying the use of digital technologies for core business functions of organisations, is serious.

Summary

The chapter presented the concept behind the need to implement digital technologies for FM processes. Some selected digital technologies for FM were reviewed with their potential benefits outlined. The digital technologies shown are computer-aided facilities management, Building Information Modelling, the Internet of Things, Geographic Information Systems, and virtual reality. Furthermore, the chapter highlighted some of the challenges to the implementation of digital technologies in effective FM processes. On a concluding note, it is emphasised that these digital technologies can help resolve some of the challenges attributed to the use of traditional methods of FM delivery.

References

Abdullah, S.A., Sulaiman, N., Latiffi, A.A., and Baldry, D. (2014). *Building information modelling (BIM) from the perspective of facilities management (FM) in Malaysia*. International Real Estate Research Symposium 2014 (IRERS), Ministry of Finance, Malaysia, 29–30 April 2014.

Affia, I. and Aamer, A. (2022). An Internet of Things-based smart warehouse infrastructure: design and application. *Journal of Science and Technology Policy and Management*, 13(1), 90–109.

Akcamete, A., Akinci, B., and Garrett Jr., J. (2010). *Potential utilisation of building models for planning maintenance activities*. In Proceedings of the International Conference on Computing in Civil and Building Engineering (ICCCBE), Nottingham, 30 June–2 July 2010.

Amano, K., and Lou, E. (2016). BIM for existing facilities: Feasibility of spectral image integration to 3D point cloud data. *MATEC Web of Conferences*, 66(24), 1–6.

Araszkiewicz, K. (2017). *Digital technologies in facility management – the state of practice and research challenges.* Creative Construction Conference, Procedia Engineering, Primosten, Croatia, 19–22 June 2017.

Araya, D., Grolinger, K., El Yamany, H., Capretz, M., and Bitsuamlak, G. (2017). An ensemble learning framework for anomaly detection in building energy consumption. *Energy and Buildings*, 144, 191–206.

Arayici, Y. (2008). Towards building information modelling for existing structures, *Structural Survey*, 26(3), 210–222.

Asare, K., Issa, R., Liu, R., and Anumba, C. (2021). BIM for facilities management: Potential legal issues and opportunities. *Journal of Legal Affairs and Dispute Resolution in Engineering and Construction*, 13(4).

Atkin, B. and Brooks, A. (2009). *Total Facilities Management*. Chichester: Wiley-Blackwell.

Azhar, S., Khalfan, M., and Maqsood, T. (2012). Building information modelling (BIM): now and beyond. *Australasian Journal of Construction, Economics and Building*, 12(4), 15–28.

Aziz, N., Nawawi, A., and Ariff, N. (2016). ICT evolution in facilities management (FM): Building Information Modelling (BIM) as the latest technology. *Procedia – Social and Behavioral Sciences*, 234, 363–371.

Becerik-Gerher, B., Jazizadeh, F., Li, N., and Calis, G. (2012). Application areas and data requirements for BIM-enabled facilities management. *Journal of Construction, Engineering and Management*, 138(3), 431–442.

BIM Forum (2015). Level of development specification for building information models, version 2015.

Boukamp, F. and Akinci, B. (2007). Automatic processing of construction specifications to support inspection and quality control. *Automation in Construction*, 17, 90–106.

Building Information Modelling (BIM) Task Group (2013). Welcome to the BIM Task Group website. Available at: http://www.bimtaskgroup.org/

Chen, C., Dib, H., and Lasker, G. (2011). *Benefits of implementing building information modelling for healthcare facility commissioning*. In Proceedings of the ASCE International Workshop on Computing in Civil Engineering, pp. 578–585.

CRC Construction Innovation (2007). *Adopting BIM for Facilities Management: Solutions for Managing the Sydney Opera House.* Brisbane, Australia: Cooperative Research Centre for Construction Innovation.

East, B. (2013). The COBie guide. The National Institute of Building Sciences. Available at: http://www.nibs.org/?page=bsa_cobieguide (accessed 5 July 2021).

Eastman, C., Teicholz, P., Sacks, R., and Liston, K. (2011). *BIM Handbook: A Guide to Building Information Modelling for Owners, Managers, Designers, Engineers and Contractors*, 2nd edn. New York: John Wiley & Sons Inc.

Ergen, E., Akinci, B., East, B., and Kirby, J. (2007). Tracking components and maintenance history within a facility utilizing radio frequency identification technology. *Journal of Computing in Civil Engineering*, 21, 11–20.

Fathi, H. and Brilakis, I. (2011). Automated sparse 3D point cloud generation of infrastructure using its distinctive visual features. *Advanced Engineering Informatics*, 25(4), 760–770.

Gnanarednam, M. and Jayasena, H.S. (2013). *Ability of BIM to satisfy CAFM information requirements.* Paper presented at The Second World Construction Symposium 2013: Socio-Economic Sustainability in Construction, Colombo, Sri Lanka.

Gu, N., Singh, V., London, K., Brankovic, L., and Taylor, C. (2008). *Adopting building information modelling (BIM) as collaboration platform in the design industry*. Paper presented at CAADRIA 2008: Beyond Computer-Aided Design: Proceedings of the 13th Conference on Computer Aided Architectural Design Research in Asia, Chiang Mai, Thailand, 9–12 April 2008.

Gubbi, J., Buyya, R., Marusic, S., and Palaniswami, M. (2013). Internet of Things (IoT): A vision, architectural elements, and future directions. *Future Generation Computer Systems*, 29, 1645–1660.

Hichri, N., Stefani, C., De Luca, L., Veron, P., and Hamon, G. (2013). From point cloud to BIM: A survey of existing approaches. *International Archives of the Photogrammetry, Remote Sensing and Spatial Information Sciences*, XL-5/W2, 343–348.

Jung, J., Hong, S., Jeong, S., Kim, S., Cho, H., Hong, S. and Heo, J. (2014). Productive modelling for development of as-built BIM of existing indoor structures. *Automation in Construction*, 42, 68–77.

Kang, T. and Hong, C. (2015). A study on software architecture for effective BIM/GIS-based facility management data integration. *Automation in Construction*, 54, 25–38.

Kang, T.W., Park, S.H., and Hong, C.H. (2016). *BIM/GIS-based data integration framework for facility management*. Paper presented at GEO Processing 2016: The Eighth International Conference on Advanced Geographic Information Systems, Applications, and Services.

Karan, E.P. and Irizarry, J. (2014). Developing a spatial data framework for facility management supply chains. In D. Castro-Lacouture, J. Irizarry, and B. Ashuri (eds), *Construction Research Congress 2014*. Reston, VA: ASCE, pp. 2355–2364.

Kassem, M., Kelly, G., Dawood, N., Serginson, M., and Lockley, S. (2015). BIM in facilities management application: A case study of a large university complex. *Built Environment and Project Asset Management*, 5(3), 261–277.

Kelly, G., Serginson, M., Lockley, S., Dawood, N., and Kassem, M. (2013). BIM for facility management: A review and a case study investigating the value and challenges. In N. Dawood and M. Kassem (eds), *Proceedings of the 13th International Conference on Construction Applications of Virtual Reality*.

Kim, J.E. and Hong, C.H. (2017). A study on facility management application scenario of BIM/GIS modelling data. *International Journal of Engineering Science Invention*, 6(11), 40–45.

Klein, L., Li, N., and Becerik-Gerber, B. (2012). Imaged-based verification of as-built documentation of operational buildings. *Automation in Construction*, 21, 161–171.

Ko, C.H. (2017). *RFID, web-based, and artificial intelligence integration in facilities management*. Paper presented at 34th International Symposium on Automation and Robotics in Construction (ISARC 2017).

Lee, J., Jeong, Y., Oh, Y., Lee, J., Ahn, N., Lee, J., and Yoon, S. (2013). An integrated approach to intelligent urban facilities management for realtime emergency response. *Automation in Construction*, 30, 256–264.

Lin, Y.C., Cheung, W.F., and Siao, F.C. (2014). Developing mobile 2D barcode/RFID-based maintenance management system. *Automation in Construction*, 37, 110–121.

Liu, R. and Issa, R. (2016). Survey: Common knowledge in BIM for facility maintenance. *Journal of Performance of Constructed Facilities*, 30(3), 04015033.

Love, P.E., Simpson, I., Hill, A., and Standing, C. (2013). From justification to evaluation: building information modelling for asset owners. *Automation in Construction*, 35, 208–216.

Meerow, S., Newell, J.P., and Stults, M. (2016). Defining urban resilience: A review. *Landscape and Urban Planning*, 147, 38–49.

Melfi, R., Rosenblum, B., Nordman, B., and Christensen, K. (2011). *Measuring building occupancy using existing network infrastructure*. Paper presented at The Green Computing Conference and Workshops (IGCC), 2011 International, Orlando, FL, USA, 25–28 July 2011.

Motamedi, M. and Hammad, S.A. (2013). Localisation of RFID-equipped assets during the operation phase of facilities. *Advanced Engineering Informatics*, 27(4), 566–579.

Newsham, G.R., Xue, H., Arsenault, C., Valdes, J.J., Burns, G.J., Scarlett, E., Kruithof, S.G., and Shen, W. (2016). Testing the accuracy of low-cost data streams for determining single-person office occupancy and their use for energy reduction of building services. *Energy and Buildings*, 135, 137–147.

Oti, A.H., Kurul, E., Cheung, F., and Tah, J.H. (2016). A framework for the utilisation of building management system data in building formation models for building design and operation. *Automation in Construction*, 72, 195–210.

Rycroft, M. (2006). Optimal use of facilities: Free State's strategic approach. *Journal of Facilities Management*, 22, 32–39.

Sabol, L. (2008). *Building information modelling & facility management: The power of process in the built environment*. Paper presented at IFMA World Workplace, https://pdfs.semanticscholar.org/cb40/288c07f9351120447f872a2e815f5f1a8db5.pdf (accessed 10 July 2021).

Shen, L., Edirisinghe, R., and Yang, M.G. (2016). *An investigation of BIM readiness of owners and facility managers in Singapore: institutional case study*. In Proceedings of the CIB World Building Congress, Vol. IV: Understanding Impacts and Functioning of Different Solutions, pp. 259–270.

Sidek, N., Ali, N., and Alkawsi, G. (2022). An integrated success model of Internet of Things (IoT)-based services in facilities management for public sector. *Sensors*, 22(9), 3207.

Succar, B., Saleeb, N., and Sher, W. (2016). *Model uses: Foundations for a modular requirements clarification language*. Paper presented at Australasian Universities Building Education (AUBEA2016), Cairns, Australia, 6–8 July 2016.

Teicholz, P. (2013). *BIM for Facility Managers*, 1st edn. Hoboken, NJ: John Wiley & Sons.

Terreno, S., Anumba, C.J., Gannon, E., and Dubler, C. (2015). *The benefits of BIM integration with facilities management: A preliminary case study*. Paper presented at 2015 International Workshop on Computing in Civil Engineering.

Volk, R., Stengel, J., and Schultmann, F. (2014). Building information modelling (BIM) for existing buildings: Literature review and future needs. *Automation in Construction*, 38, 109–127.

Wu, Q., Ding, G., Xu, Y., Feng, S., Du, Z., Wang, J., and Long, K. (2014). Cognitive internet of things: A new paradigm beyond connection. *IEEE Internet of Things Journal*, 1(2), 129–143.

Xiong, X., Adan, A., Akinci, B., and Huber, D. (2013). Automatic creation of semantically rich 3D building models from laser scanner data. *Automation in Construction*, 31, 325–337.

Part III

Cyber-physical systems

4 Theoretical perspectives on cyber-physical systems

Introduction

One of the distinct modern revolutionary technological applications in various spheres of human endeavours is the advent of cyber-physical systems (CPS). A cyber-physical system is a mechanism that integrates computation and physical processes. It is deployed through the monitoring of computer-based algorithms connected to the internet and serves as the link between the physical and virtual worlds. The defining features of CPS are the cyber capability of physical components, a high level of automation, multiple scale networking, spatial and temporal integration at multiple scales and re-assembling dynamics. In the case of construction works, virtual models are designed as a prototype of the envisaged built facility and when the facility is completed, as-built drawings are designed to give a true reflection of the various components that make up the finished facility. One primary goal for creating the as-built drawings is to facilitate easy and convenient management of the completed structure with a prominent emphasis on the maintenance of the facility. The integration of CPS into the built facility, in conjunction with the as-built virtual models, aids in seamless management of the facility. This set-up can encourage real-time computation of the state of the components installed in the facility. If any component in the facility malfunctions, a bi-directional coordination signal between the physical component and the virtual model, synchronised by the computational ability of the CPS, would be sent to alert the facilities manager to prompt correctional measures.

Overview of cyber-physical systems

According to the National Science Foundation (2013):

> CPS are systems where physical and software components are deeply intertwined, each operating on different spatial and temporal scales, exhibiting multiple and distinct behavioural modalities, and interacting with each other in a myriad of ways that change with context.

DOI: 10.1201/9781003376262-7

A cyber-physical system is a system that connects and controls the interaction between physical and computational systems or informatics (including communication) in a compact operation (Wu and Li, 2012). Gunes et al. (2014) define CPS as a complex, multi-disciplinary, physically-aware next-generation engineered system that integrates embedded computing technology (the cyber part) into the physical phenomena by using transformative research approaches. This integration mainly includes observation, communication, and control aspects of the physical systems from the multi-disciplinary perspective. In this book, CPS are defined as a computational mechanism premised on the fusion of the physical and cyber worlds, whereby there is a bi-directional exchange of information between both environments.

Baheti and Gill (2011a) noted that CPS could be viewed as transformative technologies to manage interlinked systems involving computational capabilities and physical assets. Furthermore, MacDougall (2014) noted that CPS consist of enabling technologies that integrate the physical and virtual worlds to create a wholly networked environment, whereby the interaction and communication between intelligent objects can be carried out. Meanwhile, a cyber-physical system can be portrayed as a system that connects the physical world with the world of computing and communication (Lee, 2015). Lee advocates that CPS are mainly centred on the intersection, not necessarily the cyber and physical union. CPS merge engineering methods and models from mechanical, industrial, aeronautical, chemical, bio-medical, electrical, civil, and environmental engineering with methods and models of computer science (Lee, 2008). In a typical cyber-physical system, operations are integrated, controlled, coordinated, and monitored using computing and communications (Hu et al., 2012). It can be viewed as part of how changes were introduced to human interaction by the internet. CPS bring about changes in the interaction and control between humans and the physical world. To summarize, CPS integrate real-time systems, embedded systems, allocated sensor systems, and controls (Kim et al., 2021). Emphasis is on complex mutual reliance and connection of the physical world and cyberspace characterised by close integrated communication, computation, physical elements, and control.

According to Lee (2015), CPS comprise two main functional components: (1) a real-time collection of data through advanced connectivity between the physical world and cyberspace information feedback; and (2) computational constructs and intelligent management of data and the analytics that formulates cyberspace. Furthermore, Wu et al. (2011) noted that CPS connect the cyber environment (communication, information, and intelligence) to the physical environment by sensors. This is further supported by Lee (2008), who defines CPS as the conjunction of physical and computational processes which entails control and computation units, communication web, sensors, and actuators. According to Yen et al. (2014), the fundamental components comprising a cyber-physical system are sensor networks, embedded systems, and cloud platforms. No matter the CPS context, the basic components are the same. According to this theory, CPS can be described as the interconnection between distributed sensor systems and control, real-time systems, and embedded systems (MacDougall, 2014). Lee (2015) states that CPS have features, including embedded and mobile sensing; and cross-domain sensor and data flow.

The architecture of a cyber-physical system comprises four layers (Wang et al., 2012): (1) service implementation; (2) service abstraction; (3) business process; and (4) application layers. The layer that functions as the foundation of the system's architecture is the service implementation layer. This consists of sub-categories, namely the sense-actuation unit, the communication unit, and the computation-control unit. Monitored information elicited from the physical world is captured by the sensor unit and subsequently transferred to the computation unit via the communication unit. Subsequently, the actuation unit receives instructions from the control unit for physical process control. Sensors and actuators linked with the terminal computation module make up the sense-actuate unit. Invariably, physical processes are monitored by sensors, while the actuators control the computational processes. The guiding principle of storing real-time data forms the terminal computation module. The communication mechanism makes up the communication unit using 3G and 4G systems, among others. Discrete and continuous domains are connected by the computation unit, while the control unit executes the management of space and time of a system. Generally, the main defining features of a cyber-physical system are a multiple scaling network, component cyber capability, a high level of automation, re-assembling dynamics and temporal integration at multiple scales (Lee, 2015).

There are two basic distinct functioning modes of CPS: continuous and discrete functions. The interwoven functions in CPS can be referred to as a two-state "on" and "off" functions (Knight et al., 2017). Also, the discrete functions are performed by switches in the physical world; however, the signals transmitted in the physical world to change the state of logic function are continuous. The signals generated or signals reacted to by logic function are discrete. Time is a vital element that distinguishes between continuous and discrete functions.

Models and analyses of CPS

CPS are usually engineered using models (Hu et al., 2016). Models are a formal representation depiction of a system or the components making it up (Rothenberg et al., 1989). A typical example is a model of a cyber-physical system that depicts linear hybrid automation (LHA). This is characterised by a well-defined mathematical layout that connects discrete jumps and uninterrupted evolutions (Peng et al., 2020). The semantics of the model, which also is the model's meaning, are laid out in a formal mathematical structure. From the perspective of engineering, the decisive source of information is the semantics of the model. The semantics are expressed as behaviours permitted by the model. Behaviours are dependent on the time, computation, and concept of the model; hence the behavioural features of CPS models are heterogeneous (Fitzgerald et al., 2015). The model's specifications are depicted using *formalisms* (Broman et al., 2012). This represents languages accompanied by a formal syntax that forms the basis of explicit rules to produce sentences in a language. The syntax is mapped to the semantics to provide meaning to the language. Furthermore, each model possesses a *referent* which is the component of the system intended to be represented. Usually, a system having multiple models often experiences a partial overlap of referents, leading to a description in multiple folds of the same system parts. The applicability of some analyses is enabled in a certain *context* – the given model's "condition" that is created by the analysis. The logical restraints depict the condition of the model. When a check is carried out prior to executing the analyses, these restraints are termed *assumptions*. A typical example is an assumption in checking thread-model analyses that makes use of rate-monotonic scheduling (Chaki and Edmondson, 2014). Applying analyses not in the right contexts might produce false results, for example, the model might report a defective system as being appropriate. After the execution of the analysis, which is devoid of errors, the model conditions are called the *guarantees* of the analysis.

Components of cyber-physical systems

According to Baheti and Gill (2011b), researchers have made remarkable progress in developing tools and methods in the systems and control domains regarding the frequency and time domain methods, filtering, state-space analysis, robust control, optimisation, system identification, and stochastic control. Also, there have been major research breakthroughs in computer science in visualisation methods, programming languages, embedded systems, techniques for real-time computing, compiler designs, and innovative approaches to ensure the reliability of computer systems, fault tolerance and cyber security. Furthermore, a wide range of verification tools and powerful modelling formalisms by computer science researchers have been developed. The aim of research on CPS is to combine engineering principles and knowledge within the engineering and computational disciplines to further new CPS science, coupled with the necessary technology. Research that supports the theoretical and technological underpinning of CPS has various aspects, including sensing, communication, computing, cognition and autonomous control. An essential feature of CPS is sensing and communication, which serves as the interface with the physical world. It is imperative to address the sources of cross-domain sensors and the flow of data. This data sensing of the cross-domain feature is exchanged through heterogeneous networks. Furthermore, it is germane that CPS can engage in the collection, analysis, processing, and reaction to the various types of sensing, communications, including other types of data captured during the service operations (Kao et al., 2015). However, designing and implementing connected systems face many challenges associated with variable time delays, event-driven computing, computer failures, failures in software, decision support systems, or reconfiguration (Baheti and Gill, 2011b).

The part of the information systems that handles the increasing information provided is the CPS; this makes it difficult to keep humans in the loop regarding the pertinent decisions to react to in a given situation. Furthermore, CPS are considered essential in core safety situations since the system offers services that conventional technology cannot deliver (Kao et al., 2015). This requires using CPS to imagine infrequent scenarios in order to evaluate the possibility of fatalities. Thus, this calls for a comprehensive cerebral attitude in different circumstances (Poovendran, 2010). Some basic problems, which include intelligent localisation, clock synchronisation, and adaptive sensing, provide the intelligence in the embedded systems that should be appropriately assessed by factoring in conscious and automatic intelligent mechanisms (Alippi, 2014). The possibility of computers becoming ubiquitous, also known as autonomic computing, is another important aspect. A complex computing system, such as a cyber-physical system, involves an enormous connection of computing devices, coupled with smart objects, attributed with multiple and complex interactions, thus creating more difficulties in its management. This means that autonomic computing accommodates the basics of the systems that are fundamental to the system's functioning, with little user input, hence no need for reprogramming. Ribeiro et al. (2010) noted that the CPS' components are modules made up of fine granularity incorporating self-enabling computational abilities.

Furthermore, of note is that in the specifications and definitions of CPS, it is vital to consider the intrinsically concurrent features of CPS, i.e., the coordination of the physical and cyber subsystems. A typical model of CPS entails the models of physical systems working in conjunction with computational systems, software, and networks. According to Derler et al. (2012), the bi-directional synergy between the physical systems and the computation involves sensors, actuators, software scheduling, computation, physical dynamics, and networks with delays in communication and contention. It is quite challenging to model such systems, and it requires the synchronisation of networks, software engineering, control engineering, and sensors (ibid.). A peculiar feature attributed to CPS is the nonlinear interaction physical sub-layer (continuous phenomena) and computing device (discrete). This hugely influences research carried out on hybrid systems, which is also very important to the modelling and analysis of CPS. Moreover, another important feature is the spatial distribution of CPS and their emergent behaviours, for example, cyber-attacks, which mostly arise from the interactions of the components making up the system and are removed when specific components from the systems are detached. As a result of the ubiquity and effect of CPS on different aspects of human life, one challenge is the efficient prediction of emerging behaviour of these systems. Due to the complex nature of CPS models, it is difficult to attempt to validate their safe behaviour comprehensively.

Sanislav and Miclea (2012) stated that the requirements of a cyber-physical system include an approach that is interdisciplinary or trans-disciplinary and multidisciplinary at the same time. Two major knowledge domains make up a

CPS: the physical and the cyber domains. However, the multidisciplinary attributes of CPS enable it to have more than two knowledge domains. Thus, a cyber-physical system incorporates knowledge domains, such as engineering, biological and information sciences; At the same time, the trans-disciplinary attribute includes knowledge in several domains, given its implementation and applications. This is typified in the provision of technologies and architectures in the realisation of CPS services. This is in tandem with the description provided by Pohl on trans-disciplinary research (Van der Vegte and Vroom, 2013). The adoption of modular designs is very important in attaining a robust CPS; also, the implementation methodologies are quite pertinent. This means that all the components making up the system ought to possess well-defined capabilities and interface with each other. The major challenge is the description of the sections in which the control systems of the CPS are instructed to attend to the demands for fine granularity.

Complexity theory

Complexity theory attempts to explain the modest causative factors that lead to complex behaviours (Zhu and Milanovic, 2017). All CPS systems inherently are complex systems, thus showing that they are extended spatial or temporal systems that are non-linear. A cyber-physical system is composed of numerous modules whose connection to each other creates interactions that are multi-lateral. Furthermore, the various components have a lot of freedom to act, however, they work under the limitation of functioning with global constraints in the run-time environment to attain system functions.

Multi-agent systems

In the discussion on CPS, on the infusion of a wide range of contextual information, the interfacing components within the system ought to become agents, such as humans, or a piece of software, to link sensors that then connect to computational mechanisms. According to Mittal (2012), the essential features of the agents are intelligence, an identity, and the capability to carry out functions to achieve their goals. There is a degree of freedom or autonomy on the part of the agents and the ability to collaborate or compete with others, thus being the basis for the reconfigurable orchestration to achieve the planned goals.

Semiotics

According to Galantucci et al. (2012), semiotics deals with the study of symbols or signs broadly expressed in three fundamental dimensions: (1) semantics (interlink between signs and the physical world to the sign system); (2) syntactic (operations of rules guiding the signs and the sign system); and (3) pragmatics (assessment of sign systems concerning the outlined objectives). Though originating in the biological world, this principle gives a solid theoretical basis for adoption in the application of CPS. Targon (2018) outlines the inter-link between artificial intelligence (AI) and semiotics, which is a vital and interesting basis for CPS.

Artificial intelligence (AI) and reasoning

Over the years, artificial intelligence has been the basis for research and studies of machine learning principles and algorithms. Numerous reasoning techniques have also been advanced to represent methodologies on formal knowledge. One prominent tool is KEEL (Knowledge Extraction based on Evolutionary Learning), which is software that aids the management of data. Also, it has the capability to perform evolutionary learning concerning different challenges in data mining, such as unsupervised learning, classification, and regression (Mittal, 2012).

Design of cyber-physical systems

A cyber-physical system is a fusion of systems characterised as complex and heterogeneous, whose interaction is exhibited in continuous mode. Also, the articulate regulation of the design of the system's general architecture is vital, which is prompted by the need for seamless delivery of the working modes of CPS. To achieve this, different studies have outlined different designs, modelling techniques, and tools for programming CPS.

The architecture of CPS

Abstraction made by models can be useful to portray physical realities with ideal features such as determinism. In order to have confidence in the physical set-up of systems, high-fidelity models are important. Due to the heterogeneous nature of CPS, problems are usually encountered during modelling; furthermore, these challenges result from real-time requirements and varying physical processes concurrently. These abstractions differ due to granularity, their underlying physics, and the details needed. CPS architectures and frameworks appear to be inadequate, as a result of the system's heterogeneous mechanisms, coupled with their interactions, making the requirements for modelling quite stringent. Over the years, various modelling approaches to the design of CPS have emerged. These were spurred by the necessity to address the heterogeneous nature of CPS and the modelling of conceptualised systems, whose requirements are application-specific. Various researchers have deployed meta-programming and meta-modelling techniques, such as adopting various semantics methodologies, which include operational, axiomatic, denotational, or a hybrid of these. Furthermore, different models are used in the representation of CPS, which include the actor-oriented approach, activity-centred semantic models, and multi-agent semantic models (Targon, 2018).

The importance of the various architectural and programming languages used in the modelling of the attributes of the physical and cyber components making up a cyber-physical system cannot be over-emphasised, also important is the need for methodologies in the specification and verification of properties of the different components making up the system. Hang et al. (2011) proposed different techniques that would enable the synthesis of the algorithm of cyber-physical models of architecture that face constraints with real-time attributes. Language with meta-architectural specification was adopted. This enables the designers to give specifications on the needed attributes of the architectural model, even without specifying how these can be achieved. Moreover, the study developed modulo theories, based on integer linear programming, possessing a scheduling theory solver. Further, it adopted the synchronisation of cyber-physical architectural models' constraints, such as real-time attributes. The application of this system has been tested in industrial designs with large-scale characteristics. Rajhans et al. (2009) presented a methodological meta-architectural design to facilitate the design and evaluation of substitute CPS' architecture, using Acme architectural description language; thus extending the prevailing architectural description languages, such as the Abstract Architecture Description Language (AADL), the Systems Modeling Language (SysML), and the Unified Modeling Language (UML), by permitting component modelling using formal methods. The formal methods include hybrid automata, labelled transition systems, and finite-state processes, with the use of components' behaviours. This method uses plug-ins to carry out behaviour analysis. Furthermore, the study gives a basis for a unified framework that models the elements of the physical and cyber aspects of CPS.

Dabholkar and Gokhale (2009) outlined a method that systematically specialises in general-purpose middleware in attaining the demands of CPS used in various domains. This method is based on the principles of feature-oriented software by employing an algebraic arrangement of existing middleware. This method uses origami matrices in the modelling of combined features, provided by the system's components. This makes it more expressive when compared to binary decision matrices. Bujorianu and Barringer (2009) developed a framework for multi-agent model CPS that served as a reference model and a formal logic used to portray CPS' safety properties. The agents making up the framework are characterised by continuous physical mobility and equally developing in a physical environment that is uncertain, diligently modelling the real world. Furthermore, the framework models, both automated control and human control, enable the user-centric model. Talcott (2008) advanced the development of semantics for CPS that is event-based. The study classified various events based on their features. It was observed that the adoption of semantics that is event-based offers a natural way to specify the components of open systems concerning observable behaviour and terms of interfaces. Moreover, this serves as a basis for designing, monitoring and implementing CPS. Bujorianu and Barringer (2009) presented a formal method, termed the

Hibertean formal method, formulated to provide denotational semantics using an algebraic model to model observability and physical causality by physical processes.

The CPS programming framework

Many frameworks and tools have been developed to aid the seamless implementation and deployment of CPS tools. One such approach is synthesis: this supports converting some form of CPS specifications into implementation. Martin and Egerstedt (2012) outlined the use of system tools to translate the CPS's high-level specification into real execution and subsequent implementation in physical devices. In bridging the gap between actual implementation and high-level specification, the study made specifications for CPS concerning motion description languages. This leads to a reduction in the complex nature of implementation and permits the structuring of tasks into units that the system can easily interpret. Roy et al. (2011) presented a design tool for CPS, termed PESSOA. The tool synthesises CPS's controllers and accepts CPS portrayed as automata and a group of differential equations. It outputs a system controller that implements the outlined specification to the given abstraction parameters. Also, another aspect is the automation of the deployed CPS programs. Hnat et al. (2008) worked on a macro-programming framework, termed MacroLab. With the use of this framework, a single program can be written for the whole CPS, while the program can be broken down into groups as micro-programs and subsequently slotted to fill individual nodes. The decomposition of the macro-program by MacroLab is done in a manner that is suited for a designated deployment. Data manipulation by the program is permitted using sensors and actuators.

Model testing and CPS verification

Several studies have attempted to verify the appropriateness of the design and the relevant properties of CPS. Bhave et al. (2011) made a proposal based on an approach for assessing and defining the consistency among architectural projections from various models that are heterogeneous. The study formulated the constraint of consistency verification as a typed graph, comparing constraints between the system's base architecture and the connectivity graph of the various architectural views. Also, a pertinent aspect is the correct working conditions of the system at its aggregate level, as against the working conditions of the system's components at its granularity level. Sun et al. (2007) checked the model to verify the appropriateness of the configuration in a power grid CPS, with the assumption of a good working condition of the individual components. With the adoption of a decomposition method, the system is divided into smaller modules, and verification of these can be done efficiently. McMillin and Akella (2011) verified CPS's confidentiality properties. As a result of the nature of CPS, an observer can infer the sensitivity of cyber actions on the flow of physical information. A typical example is the operations in a wind turbine that hugely depends on wind velocity and physical size. These are observable; these properties can reveal the cyber attributes of the system. The study presented an approach for the verification of information flow to reduce these constraints.

Applications of cyber-physical systems

CPS have been applied in various fields of human endeavours. This includes agriculture, banking, education, transportation and automobiles, energy management, environmental monitoring, health and medical systems, manufacturing and process control, security and surveillance, smart city, and construction. The application of CPS in the agricultural processes can boost food production and increase crop management efficiency through technological innovations, such as intelligent water management, precision agriculture, and efficient food distribution. Through constant monitoring of the prevailing environmental impacts on crops, the maximum agricultural output can be achieved. Caramihai and Dumitrache (2015) outlined a control procedure for CPS that guarantees a proactive agricultural system against market and environmental changes. This was based on the fact that the rise in food demand-supply gulf can be adequately discussed by real-time control services applied in agriculture.

In championing the course for road safety, Loos et al. (2011) formulated an automobile control system with vehicle navigation optimisation that aims to reduce or avoid collisions. The optimisation of the application of CPS in road safety would promote a great reduction in motorway accidents. Furthermore, with the use of a wireless sensor network (WSN), Yan et al. (2012) developed a model to optimise system delivery of transportation modes for unmanned CPS. Equally, this model seeks to enhance automobile navigation to reduce the likelihood of accidents for road users.

For efficient medical care, the application of CPS in the processes and administration of health care delivery has long been accepted. Milenkovic et al. (2006) observed that technological innovations, such as a wireless sensor network (WSN), medical sensors, and cloud computing have established CPS as a viable candidate for efficient medical applications in both hospital and home patient care. Likewise, for proper and adequate energy management systems, CPS are very useful. Li et al. (2016) designed a distributed model with data attack for a class of power systems. The model adopts a 9-bus power system of state estimators used to control distributed systems with large-scale optimisation.

Lee (2015) stated that the integration of CPS in production, services, logistics, and manufacturing systems would transform the present factories into Industry 4.0 factories with comparative advantages. Meeting the demands of the Fourth Industrial Revolution can rapidly be enhanced by combining production and process systems with a sophisticated, intelligent network. The application of CPS in production systems was discussed by

Wiesner et al. (2017). Emphasis was placed on the multidisciplinary requirement for the software, hardware, and service constituent to be prominent aspects in the successful adoption of CPS in production systems.

In the area of security and surveillance, CPS have greatly aided in helping in crime reduction. This has been through establishing a nexus-based system adopting cyber connections via general packet radio codes using various access (Ma et al., 2010).

For smart homes and smart cities, CPS present great opportunities for innovative applications through a wide range of smart building appliances. According to Cassandras (2016), a smart city translates to an urban environment encompassing new generation frontiers of new services for environmental monitoring, transportation, health care, energy distribution, emergency response, business, and social activities. Li et al. (2010) developed a modelled cyber-physical system incorporating architecture established on smart community architecture, called Networked homes. In this context, activities such as switching on electrical appliances or heating system control or deciding when to pull down curtains or security architecture are all enabled in a CPS-enabled platform. Furthermore, Seiger et al. (2016) proposed a CPS-enabled smart city home domain that is user-friendly and controlled by mobile phones. The control ability offers a drastic reduction in the difficulty experienced in using CPS with the provision of a wide range of control options, enabling non-expert users to have full access to components, such as sensors, actuators, and other complex constituents represented by service robots.

Cyber-physical systems for facilities management

Facilities management is plagued with a wide range of problems which include detection time of dysfunctional components, rise in the cost of energy, discontent on the part of occupants over allocated spaces, etc. (Han et al., 2012). Several FM systems have passive attributes with inherent parameters that are pre-programmed, while they are not formulated to accommodate flexible, complicated, and altering scenarios. The task of surmounting some of the problems has led to the formation of systems that help in the efficient and effective capturing and assessment of the concerted data flow, involving the end users of the facility. Introducing innovative technologies for improved monitoring and coordination of the management procedures associated with FM would help in delivering better FM services (Terreno et al., 2020). A CPS approach to deliver FM mandates would enhance the coordination and monitoring of the constituents of a facility, aided with functionalities that are in real time, resulting from the amalgamation of physical and virtual environments (Akanmu and Anumba, 2015). Also, the concept of FM driven by cyber functionalities can stimulate the organisation of FM to engage in cognitive functions, hence, leading to the identification of the physical environment premised on a high-level intelligence (Wu et al., 2014). This is achieved by assembling real-time sensory data in the various constituents of the facility, abetted by WSNs. Ideally, operational control, computation, and communication will form the base on which the system relies.

The nature of FM is trans-disciplinary or even multidisciplinary, entailing knowledge from a spectrum of professions, which includes design, management, engineering, accounting, architecture, finance, and behavioural science (Teicholz, 2001). These functions are closely linked to the input of human efforts and coordination in attaining the targeted goal of the organisation. However, the typical CPS do not deliver a quantifiable estimation of the influence or effects of human contributions and control. This is due to the complexity, diversity, and uncertainty associated with the contribution of humans to the delivery of FM tasks. The operations of humans in physical systems are not accounted for in the functionalities of CPS (Lee and Seshia, 2011). Analysis of the human dimension of FM can be evaluated within the purview of facility performance, within the ambits of management and control with the use of computational tools and techniques. This can be executed by interactive systems that are based on the communication between the physical and the virtual environments (Yang et al., 2018).

The concept of a cyber-driven FM would expand the structures of FM in conducting cognitive tasks, such as seeing, hearing, and smelling the physical world, thus allowing observations to be shared, while permitting the structures of FM to engage in the identification of physical worlds using a high-level intelligence (Wu et al., 2014). The assembling of sensory data in several types of facilities engaged in real time can be executed with the aid of WSNs (Huang and Mao, 2017), whereas the collection of data for behavioural and movement can be executed using photogrammetry, videogrammetry, and facility reconstruction, using digital models of as-built and as-is (Xu et al., 2020). Also, auto-IDs which include RFID and QR codes can be used in navigation, real-time tracking, and localisation in empirical cases (Xue et al., 2018). Ideally, the operation of such systems depends on elements of operational control, communication, and computation. This is aided by the four-layered CPS architecture employed for FM functions: sensing, processing, data fusion, and application (Terreno et al., 2020).

Summary

The chapter focused on the theoretical perspectives of cyber-physical systems. This presented a detailed analysis of CPS applications in different fields, their designs, modes and analyses, and components. Also, the requirements for virtual modelling of construction projects were showcased alongside CPS applications in different construction process scenarios. Also, how CPS can be applied in FM was analysed as well as the potential benefits of their use.

References

Akanmu, A. and Anumba, C. (2015). Cyber-physical systems integration of building information models and the physical construction. *Engineering, Construction and Architectural Management*, 22(5), 516–535.

Alippi, C. (2014). *Intelligence for Embedded Systems.* Cham: Springer International Publishing.

Baheti, R. and Gill, H. (2011a). Cyber-physical systems. Technical report. National Science Foundation.

Baheti, R. and Gill, H. (2011b). Cyber-physical systems. In *The Impact of Control Technology*, vol. 12. IEEE Control Systems Society, pp. 161–166.

Bhave, A., Krogh, B., Garlan, D., and Schmerl, B. (2011). *View consistency in architectures for cyber-physical systems.* In Proceedings of the 2nd IEEE/ACM International Conference on Cyber-Physical Systems, pp. 151–160.

Broman, D., Lee, E., Tripakis, S., and Torngren, M. (2012). *Viewpoints, formalisms, languages, and tools for cyber-physical systems.* Paper presented at The 6th International Workshop on Multi-Paradigm Modeling (MPM'12), Innsbruck, Austria, October 2012.

Bujorianu, M.C. and Barringer, H. (2009). *An integrated specification logic for cyber-physical systems.* In Proceedings of IEEE International Conference on Engineering and Complex Computer Systems, pp. 291–300.

Caramihai, S.I. and Dumitrache, I. (2015). *Agricultural enterprise as a complex system: A cyber physical systems approach.* In 20th International Conference on Control Systems and Computer Science (CSCS)IEEE, pp. 659–664.

Cassandras, C. (2016). Smart cities as cyber-physical social systems. *Engineering*, 2, 156–158.

Chaki, S. and Edmondson, J. (2014). Model-driven verifying compilation of synchronous distributed applications. In J. Dingel, W. Schulte, I. Ramos, S. Abraho, and E. Insfran (eds), *Model-Driven Engineering Languages and Systems: 17th International Conference, MODELS 2014*. Cham: Springer.

Dabholkar, A. and Gokhale, A. (2009). *An approach to middleware specialisation for cyber physical systems. Proceedings of the 29th IEEE International Conference on Distributed Computing Systems Workshops*, pp. 73–79.

Derler, P., Lee, E. and Vincentelli, A. (2012). Modeling cyber-physical systems. *Proceedings of the IEEE*, 100(1), 13–28.

Fitzgerald, J., Gamble, C., Larsen, P., Pierce, K., and Woodcock, J. (2015). *Cyber-physical systems design: Formal foundations, methods and integrated tool chains.* In 2015 IEEE/ACM 3rd FME Workshop on Formal Methods in Software Engineering, pp. 40–46.

Galantucci, B., Garrod, S., and Roberts, G. (2012). Experimental semiotics. *Language and Linguistics*, 6(8), 477–493.

Gunes, V.S., Peter, T., Givargis, G., and Vahid, F. (2014). A survey on concepts, applications, and challenges in cyber-physical systems. *KSII Transactions on Internet and Information Systems*, 8(12), 4242–4268.

Han, Z., Gao, R.X., and Fan, Z. (2012). *Occupancy and indoor environment quality sensing for smart buildings.* 2012 IEEE International Instrumentation and Measurement Technology Conference Proceedings, pp. 882–887.

Hang, C., Manolios, P., and Papavasileiou, V. (2011). Synthesizing cyber-physical architectural models with real-time constraints. In *Computer Aided Verification.* Berlin: Springer Verlag, pp. 441–456.

Hnat, T., Sookoor, T.I., Hooimeijer, P., Weimer, W., and Whitehouse, K. (2008). *MacroLab: A vector-based macroprogramming framework for cyberphysical systems.* In Proceedings of the 6th ACM Conference on Embedded Network Sensor Systems, pp. 225–238.

Hu, F., Lu, Y., Athanasios, Y., Vasilakos, V., Hao, Q., Ma, R., Patil, Y., Zhang, T., Lu, J., Li, X., and Xiong, A. (2016). Robust cyberphysical systems: Concept, models, and implementation. *Future Generation Computer Systems*, 56, 449–475.

Hu, L., Xie, N., Kuang, Z., and Zhao, K. (2012). *Review of cyber-physical system architecture.* In Object/Component/Service-Oriented Real-Time Distributed Computing Workshops (ISORCW), 2012 15th IEEE International Symposium, pp. 25–30.

Huang, Q. and Mao, C. (2017). Occupancy estimation in smart building using hybrid CO_2/light wireless sensor network. *Journal of Applied Sciences and Arts*, 1(2), 5.

Kao, H., Jin, W., Siegel, D., and Lee, J. (2015). A cyber physical interface for automation systems: Methodology and examples, *Machines*, 3, 93–106.

Kim, S., Eun, Y., and Park, K.-J. (2021). Stealthy sensor attack detection and real-time performance recovery for resilient CPS. *IEEE Transactions on Industrial Informatics*, 17(11), 7412–7422.

Knight, J., Xiang, J., and Sullivan, K. (2017). A rigorous definition of cyber-physical systems. In A. Romanovsky and F. Ishikwa (eds), *Trustworthy Cyber-Physical Systems Engineering*, 1st edn. New York: Chapman and Hall, pp. 48–70.

Lee, E. (2008). Cyber *physical systems: Design challenges.* Proceedings of the 11th IEEE Symposium on Object Oriented Real-Time Distributed Computing (ISORC 08), pp. 363–369.

Lee, E.A. (2015). The past, present and future of cyber-physical systems: A focus on models. *Sensors (Basel, Switzerland)*, 15(3), 4837–4869.

Lee, E.A. and Seshia, S.A. (2011). *Introduction to Embedded Systems: A Cyber-Physical Systems Approach.* California. Available at: LeeSeshia.orgLi, X., Liang, X., Shen, X., Chen, J., and Lin, X. (2010). Smart community: An Internet of Things application. *IEEE Communications Magazine*, 49(11), 68–75.

Li, Y., Wu, J., and Li, S. (2016). State estimation for distributed cyber-physical power systems under data attacks. *International Journal of Modelling, Identification and Control*, 26(4), 317–323.

Loos, S., Platzer, A., and Nistor, L. (2011). *Adaptive cruise control: Hybrid, distributed, and now formally verified.* In Proceedings of the International Symposium on Formal Methods, pp. 42–56.

Ma, L., Yuan, T., Xia, F., Xu, M., Yao, J., and Sha, M. (2010). *A high-confidence cyber-physical alarm system: Design and implementation.* In Proceedings of the IEEE/ACM International Conference on Green Computing and Communications & International Conference on Cyber, Physical and Social Computing, pp. 516–520.

MacDougall, W. (2014). *Industrie 4.0: Smart Manufacturing for the Future.* Germany: Trade & Invest.

Martin, P. and Egerstedt, M. (2012). Hybrid systems tools for compiling controllers for cyber-physical systems. *Discrete Event Dynamic System*, 22(1), 101–119.

McMillin, B. and Akella, R. (2011). *Verification of information flow properties in cyber-physical systems.* In Proceedings of Workshop FDSCPS, 2011, p. 37.

Milenkovic, A., Otto, C., and Jovanov, E. (2006). Wireless sensor networks for personal health monitoring: issues and an implementation. *Computer Communications*, 29(13), 2521–2533.

Mittal, P. (2012). Knowledge Extraction based on Evolutionary Learning (KEEL): Analysis of development method, genetic fuzzy system. *International Journal of Computer Applications and Information Technology*, 1(1), 22–25.

National Science Foundation (NSF). (2013). Cyber physical systems. NSF10515. Arlington, VA: NSF. Available at: http://www.nsf.gov/pubs/2010/nsf10515/nsf10515.htm (accessed 22 July 2020).

Peng, H., Liu, C., Zhao, D., Hu, Z., and Han, J. (2020). Security evaluation under different exchange strategies based on heterogeneous CPS model in interdependent sensor networks. *Sensors*, 20(21), 6123.

Poovendran, R. (2010). Cyber-physical systems: Close encounters between two parallel worlds. [Point of View]. *Proceedings of IEEE*, 98, 1363–1366.

Rajhans, A., Cheng, S., Schmerl, B., Garlan, D., Krogh, B., Agbi, C., and Bhave, A. (2009). *An architectural approach to the design and analysis of cyber-physical systems.* In Proceedings of the 3rd International Workshop on Multi-Paradigm Modeling (MPM 2009), *pp.* 1–10.

Ribeiro, L., Barata, J., Cândido, G., and Onori, M. (2010). Evolvable production systems: An integrated view on recent developments. In G.Q. Huang, K.L. Mak, and P.G. Maropoulos (eds), *Proceedings of 6th CIRP-Sponsored International Conference on Digital Enterprise Technology.* Berlin: Springer, pp. 841–854.

Rothenberg, J., Widman, L., Loparo, K., and Nielsen, R. (1989). *The Nature of Modeling: Artificial Intelligence, Simulation and Modeling*. Hoboken, NJ: John Wiley & Sons, pp. 75–92.

Roy, P., Tabuada, R., and Majumdar, R. (2011). *Pessoa 2.0: A controller synthesis tool for cyber-physical systems.* In Proceedings of the International Conference on Hybrid Systems: Computation and Control, pp. 315–316.

Sanislav, T. and Miclea, L. (2012). Cyber-physical systems: Concept, challenges and research areas. *Journal of Control Engineering Applications and Informatics*, 14, 28–33.

Seiger, R., Struwe, S., Lemme, D., and Schlegel, T. (2016). *An interactive mobile control center for cyber-physical systems.* In Proceedings of the ACM International Joint Conference on Pervasive and Ubiquitous Computing, pp. 193–196.

Sun, Y., McMillin, B., Liu, F., and Cape, D. (2007). *Verifying noninterference in a cyber-physical systems: The advanced electric power grid.* In Proceedings of 7th International Conference on Quality Software (QSIC 2007), pp. 363–369.

Talcott, C. (2008). Cyber-physical systems and events. In M. Wirsing, J-.P. Banâtre, M. Holzl, and A. Rauschmayer (eds), *Software-Intensive Systems and New Computing Paradigms.* Berlin: Springer, pp. 101–115.

Targon, V. (2018). Toward semiotic artificial intelligence. *Procedia Computer Science*, 145, 555–563.

Teicholz, E. (2001). *Facility Design and Management Handbook*. New York: McGraw-Hill.

Terreno, S., Akanmu, A., Anumba, C., and Olayiwola, J. (2020). Cyber-physical social systems for facilities management. In C.J. Anumba and N. Roofiari-Esfahan (eds), *Cyber-Physical Systems in the Built Environment*. Cham: Springer, pp. 297–308.

Van der Vegte, W.F. and Vroom, R.W. (2013). *Considering cognitive aspects in designing cyber-physical systems: An emerging need for transdisciplinarity.* Paper presented at International Workshop on the Future of Transdisciplinary Design (TFTD 13), Luxembourg, 24 June.

Wang, P., Xiang, Y., and Zhang, S. (2012). *Cyber-physical system components composition analysis and formal verification based on service-oriented architecture.* In 9th IEEE International Conference on e-Business Engineering, pp. 327–332.

Wiesner, S, E Marilungo and KD Thoben (2017). Cyber-physical product-service systems—challenges for requirements engineering. *International Journal of Automation Technology*, 11(1), 17–28.

Wu, F.J., Kao, Y., and Tseng, Y. (2011). From wireless sensor networks towards cyber-physical systems. *Pervasive and Mobile Computing*, 7(4). doi:10.1016/j.pmcj.2011.03.003.

Wu, N. and Li, X. (2012). RFID applications in cyber-physical systems, deploying RFID-challenges, solutions and open issues. Available at: http://www.intechopen.com/books/deploying-rfid-challenges-solutions-and-open-issues/rfid-applications-incyber-physical-system (accessed 22 July 2020).

Wu, Q., Ding, G., Xu, Y., Feng, S., Du, Z., Wang, J., and Long, K. (2014). Cognitive Internet of Things: A new paradigm beyond connection. *IEEE Internet of Things Journal*, 1(2), 129–143.

Xu, J., Lu, W., Anumba, C., and Niu, Y. (2020). From smart construction objects to cognitive facility management. In C.J. Anumba and N. Roofiari-Esfahan (eds), *Cyber-Physical Systems in the Built Environment*. Cham: Springer, pp. 273–296.

Xue, F., Chen, K., Lu, W., Niu, Y., and Huang, G. Q. (2018). Linking radio-frequency identification to building information modeling: Status quo, development trajectory and guidelines for practitioners. *Automation in Construction*, 93, 241–251.

Yan, H., Wan, J., and Suo, H. (2012). Adaptive resource management for cyber-physical systems, *Applied Mechanics and Materials*, 157/158, 747–751.

Yang, A.M., Yang, X.L., Chang, J.C., Bai, B., Kong, F.B., and Ran, Q.B. (2018). Research on a fusion scheme of cellular network and wireless sensor for cyber physical social systems. *IEEE Access*, 6, 18786–18794.

Yen, C., Liu, Y., Lin, C., Kao, C., Wen-Bin, A., and Wang Hsu, Y. (2014). *Advanced manufacturing solution to Industry 4.0 trend through sensing network and cloud computing technologies.* In IEEE International Conference on Automation Science and Engineering, pp. 18–22.

Zhu, W. and Milanovic, J. (2017). *Cyber-physical system failure analysis based on complex network theory.* In IEEE EUROCON 2017–17TH International Conference on Smart Technologies, pp. 571–575.

Part IV

Theoretical perspectives on technology adoption

5 Theoretical perspectives on technology adoption

Introduction

The adoption of technology in construction processes and the management of completed built facilities (the entire life-cycle) has been an arduous task over the years. The impact of technology in other industries, such as banking, manufacturing, and medicine, is very evident, leading to improvements in production processes and industrial systems. According to Ikuabe et al. (2022), the construction industry is still dragging its feet on adopting digital technology in its processes. With the age-long challenges posed to construction processes and delivery operations, the industry is the perfect candidate for a step change (Kamara et al., 2020). Several adoption models and theories have been proposed, including Innovation Diffusion Theory, Social Cognitive Theory, the Theory of Reasoned Action, the Theory of Planned Behaviour, the Task Technology Fit Model, Technology Acceptance Models, and the Unified Theory of Acceptance and Use of Technology (Alhassany and Faisal, 2018). A comprehensive review of these models and theories will be carried out in this book; however, the emphasis will be placed on the Unified Theory of Acceptance and Use of Technology (UTAUT) as it serves as the theoretical underpinning of the study.

Overview of technology adoption theories and models

Innovation Diffusion Theory (IDT)

The acceptance or rejection process of innovation is the basis of the Innovation Diffusion Theory (Rogers, 2003). The theory proposes five stages during decision-making for innovation. The first stage dwells on its purpose, and how it operates, it is referred to as the knowledge stage. The pertinent questions posed at this stage are: how?, what?, and why?. The next stage is called persuasion, referring to an individual's liking or dislike of new technology. According to Rogers (1995), compatibility, trialability, relative advantage, complexity, and observability are the five attributes that influence an individual's liking of new technology. Decision-making is the third stage; this stage

DOI: 10.1201/9781003376262-9

informs the individual's adoption or rejection of the technology. The fourth stage is putting the technology to use; it is termed the implementation stage. Lastly, the fifth stage centres on the individual's confirmation of the decision adopted regarding the innovation.

The Innovation Diffusion Theory (IDT) presents the way and manner of migration from invention to use of technological innovation, idea, or technique or the unfamiliar use of an old technique (Rogers, 2003). In this theory, the communication of technological innovation is conveyed through particular routes within a population, making up a social system within a given period. The underpinning concepts behind this theory seek to explain the adoption of innovation using four elements: innovation, communication channel, time, and social systems. The theory explains that in adopting new technology, an individual's behaviour is influenced by the intuitive disposition concerning relative advantage, trialability, compatibility, complexity, observability, and social norms (ibid.).

Social Cognitive Theory (SCT)

The Social Cognitive Theory of Bandura (1986) suggests that the dynamic mutual effect of personal, behaviour, and environmental forces serves as the basis for human functioning (Figure 5.1). People's interpretation of their behavioural position influences their perception of their environment and thereby the inherent factors, which predict or alter the ensuing behaviour. This serves as the basis of the idea of reciprocal determinism by Bandura (ibid.). This takes into consideration the following: individual features in the context of biological events, cognition, and effect; behaviour; and environmental conditions that engender communication that leads to triadic reciprocity.

Theory of Reasoned Action (TRA)

The Theory of Reasoned Action (TRA), a multi-lateral behavioural theory, was propounded in 1980 by Ajzen and Fishbein (Chau and Hu, 2002). The

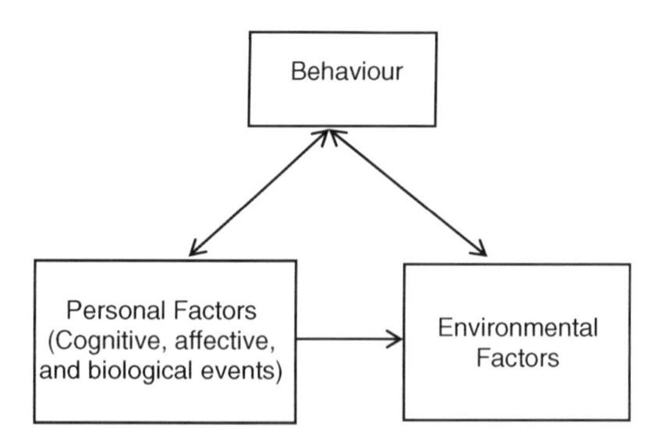

Figure 5.1 Social Cognitive Theory
Source: Bandura (1986).

model forms the backbone for studies related to the relationship between attitude and behaviour. Magee (2002) affirms that the theory has been deployed in various fields and is commonly deployed in the academic and business spheres. TRA postulates that an individual's beliefs influence their attitude and social norms, which results in adopting certain behaviours or perhaps guiding the behaviour of the individual (Ajzen and Fishbein, 1980; Leach et al., 1994). The antecedent of behaviour is an intention, which is the cognitive portrayal of an individual's willingness to act on a particular behaviour. TRA is founded on two constructs: (1) the attitude towards behaviour (ATB), and (2) the subjective norm (SN) in collaboration with that behaviour, as presented in Figure 5.2. The attitude towards the behaviour (ATB) serves as the prior attitude of an individual concerning the behaviour. People think about the decisions and possible eventualities of actions taken before taking such decisions. This theory refers to the view that a person's intention to decide on taking action or not is based on the determinant of the action. In contrast, the individual's beliefs and appraisal of the outcome determine the attitude. Subjective norm (SN) refers to the social pressure faced by an individual or what the decision-maker encounters to undertake the said behaviour. Leach et al. (1994) state that what a person thinks about other people's view on the said behaviour is referred to as an SN. In retrospect, what other people or groups think about the decision made by an individual to follow a particular behaviour and equally how these people perceive the decision both play a key role in the decision to act or not in that manner. This explains why often people consult others before making certain decisions.

Theory of Planned Behaviour (TPB)

The Theory of Planned Behaviour (TPB) is an extension of the TRA and attempts to bridge the gap in addressing the behaviour over which individuals have incomplete control. This theory brings a new phenomenon that explains

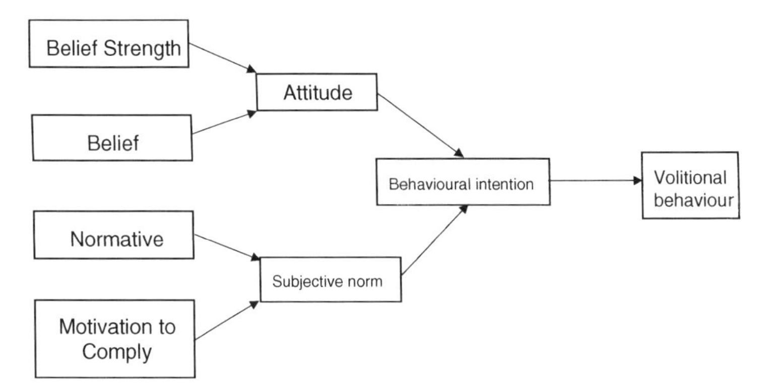

Figure 5.2 Theory of Reasoned Action
Source: Ajzen and Fishbein (1980).

an independently defined perspective called perceived behaviour control (PBC). This led to the introduction of the TPB by Ajzen in 1985 (Ajzen, 1985). The additional construct (PBC) comes into play in scenarios where the individual is not in control or lacks the resources needed to freely perform the behaviour. As shown in Figure 5.3, the theory predicts intentional behaviour, since behaviour can be planned or deliberate; with the inclusion of the PBC, the TPB is seen as more holistic than the TRA (Chau and Hu, 2002). Comparing the TRA and the TPB, both suggest that intention is the best predictor of behaviour (ibid.). For the TPB, three constructs are the foundation for intention: attitude towards a particular behaviour, subjective norms, and perceived behavioural control (Ajzen, 1991).

Yzer (2012) defined PBC as the notion of internal and external hindrances in executing actions. It refers to people's notion of their capacity to carry out a particular behaviour. In general, when an attitude and the subjective norm are more favourable, there is a high likelihood of greater control of the perceived behaviour, thus indicating the intent of the person to act in such a way. Also, Ajzen (1991) noted that specific attitudes are required to predict exact behaviours, while a general attitude is required to predict aggregated behaviours. Furthermore, a range of factors arising from an event connected to the behaviour affects the particular behaviour. The different factors in a particular behaviour will connect with each other in combination in different conditions and environments, thus implying that an inclusive attitude is the main determinant of behaviour (Davis and Songer, 2008).

According to Ajzen (2002), human behaviour is premised on three kinds of beliefs based on TPB: (1) behavioural beliefs centred on probable outcomes of the depicted behaviour and the appraisal of these outcomes; (2) normative beliefs refer to the discerned behavioural anticipation of important individuals or a certain category of persons, such as family members, co-workers, bosses, friends, or spouses. The result is subjective norms or perceived social pressure; and (3) control beliefs, which centre on the influences of a depicted act, this raises the question of the individual's control over their own behavioural disposition.

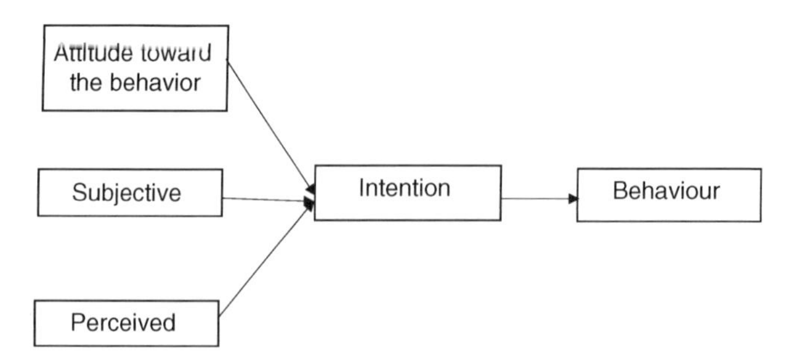

Figure 5.3 Theory of Planned Behaviour
Source: Ajzen (1991).

The Task Technology Fit (TTF) Model

This theory connects the requirements needed for technology execution with the tasks expected to be executed. The Task Technology Fit (TTF) model comprises three constructs: (1) the task characteristics; (2) the technology characteristics; and (3) the individual's characteristics (Figure 5.4). The three constructs influence the principle of "technology fit", which equally directly impacts the eventual resultant variable termed "performance". Goodhue and Thompson (1995) noted that these three constructs making up the TTF model show the technology's capability to adequately support the given task by properly incorporating the requirements of the task with the abilities of the said technology. The model proposes that technology is deployed to carry out a task, provided that the tools available to the employee are suitable to carry out such a task. When carrying out a task, experienced personnel will select the appropriate tools to facilitate the delivery and completion of the task with the desired results.

The Technology Acceptance Model (TAM)

The strength and soundness of the Technology Acceptance Model (TAM) in explaining divergent technology contexts have been affirmed by several meta-analyses (King and He, 2006). This is a generally accepted model deployed to explain system acceptance and information technology (Xu et al., 2014). Due to its narrow emphasis on beliefs related to technology's value, it is very useful and better adapted in predicting technology adoption at the individual level, compared to other adoption theories (Liu et al., 2008). TAM is an improved and modified version of the Theory of Reasoned Action (TRA) by emphasising the adoption and acceptance of technology by individuals (Lee et al., 2011). Hence, TAM seeks to outline the reasons for accepting or rejecting a given technology by an individual when undertaking a task or job (Wallace and Sheetz, 2014).

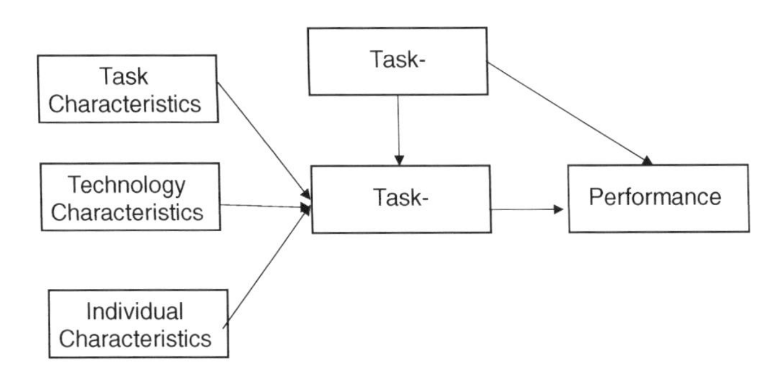

Figure 5.4 The Task Technology Fit Model
Source: Goodhue and Thompson (1995).

The concept behind TAM states that people's usage of a given technology is born out of the intention to use the technology, which is connected to feelings about the technology. Eben and Achampong (2010) stated that the TAM is based on two features that serve as determinants of the adoption of innovative technology, i.e., the perceived usefulness (PU) and the perceived ease of use (PEOU). These are deployed in the determination of the extent of the adoption of technology by individuals and organisations. The behavioural intention of adopting the technology is directly influenced by the perceived usefulness of the technology (Davis et al., 1989). TAM is an offshoot of the TRA, which is based on modelling of acceptance of information systems by users. Davis et al. (ibid.) emphasised that TAM is referred to as one of the most prominent models related to accepting, adopting, and using technology; it possesses the ability to provide explanations and predictions of users' behaviour regarding information technology. The two essential features of the model – PEOU and PU – are the basic constructs, with the PEOU having core dimensions as system design and features while the PU is based on decreasing effort (Moore, 2012). As illustrated in Figure 5.5, the theory provides the connection of attitude, belief, intention, and usage.

TAM illustrates the interconnection between the variables making up an individual's adoption and acceptance of technology. These variables eventually guide the individual's behavioural intention (BI) to use it (Autry et al., 2010). To a large extent, an individual's attitude towards use influences the individual's BI (Mathieson, 1991). The beliefs about the perceived ease of use and perceived usefulness regarding the technology directly affect the individual's attitude (Xu et al., 2014).

According to Lee et al. (2013), perceived usefulness measures a potential user's belief that using the technology will improve the job outcome or performance, from an organisational point of view. On the other hand, PEOU measures a potential user's belief that using the technology will demand less application of effort. Furthermore, TAM recognises the impact of extension

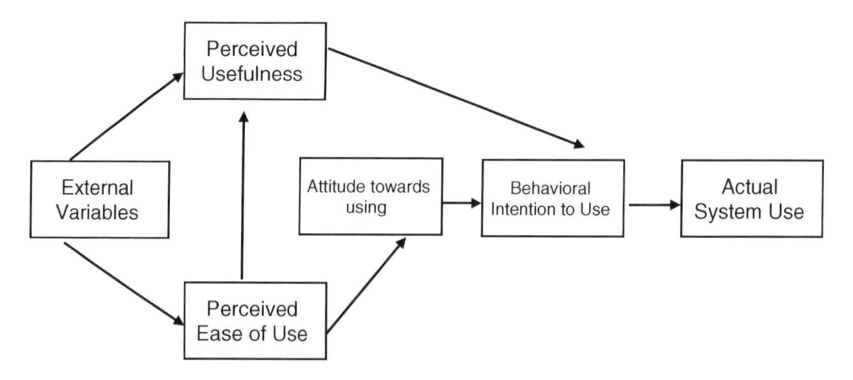

Figure 5.5 The Technology Acceptance Model (TAM)
Source: Davis et al. (1989).

variables on the intention to use technology, in which PU and PEOU serve as mediators (ibid.). The extension variables include the development process, system features, and training. Also, the theory proposes that perceived usefulness is influenced by PEOU, suggesting that technologies are easily adopted when they prove to be effortless to use (Wallace and Sheetz, 2014).

The validation of TAM has been proven and reinforced by studies in different fields, therefore, proving its reliability, application, and replication (Venkatesh and Davis, 2000). Likewise, the reliability and validity of variables making up the perceived usefulness and PEOU have been proven by different studies (Wallace and Sheetz, 2014).

The Technology Acceptance Model 2 (TAM2)

The shortcomings of TAM, such as organisational limits, limited ability, environmental factors, time, and unconscious habits, needed to be addressed. This led to Venkatesh and Davis (2000) improving on the TAM to establish the Extended Technology Acceptance Model (TAM2). This makes provision for a comprehensive interpretation of the pertinent constructs that make up the perceived usefulness. With TAM as the basis, TAM2 embodies supplementary theoretical constructs, which include useful cognitive procedures (result demonstrability, job relevance, and output quality) and social impact procedures (image, voluntariness, experience, and subjective norm), these were all absent in the initial TAM. Social impacts such as subjective norms and images were reviewed to rectify the shortcomings of the previous TAM. In real-life scenarios, restrictions such as unconscious habit, time, organisational limits, and limited ability will restrict the individual from acting freely. In comparison to the subjective norm, image is viewed as people's portrayal of how they are represented. And this, coupled with attitude, significantly impacts the perceived usefulness (Chan and Lu, 2004).

Several studies have evaluated the constructs making up the technology acceptance models, with particular emphasis on perceived usefulness. It has been judged insufficient or restricted in its application. It does not give adequate coverage of the varieties of duties involved in the user's task or the impediments of various working conditions. Legris et al. (2003) highlighted the drawbacks of TAM and TAM2 by arguing that these theories do not include variables that portray change indicators in business processes and organisational culture. Also, Chan and Lu (2004) stressed that "job effectiveness enhancement", "accomplishment of tasks quickly", and "job performance improvement" are not valid yardsticks for ascertaining perceived usefulness. Furthermore, the function of learning and reinvention concerning the measurement of perceived usefulness appears to have been overlooked (Saeed and Abdinnour-Helm, 2008).

Unified Theory of Acceptance and Use of Technology (UTAUT)

The drawbacks and shortcomings of the previously discussed adoption theories evidenced the need for an all-encompassing theory that seeks to attend to the identified lapses, e.g. the non-inclusion of vital constraints, such as needed resources (money and time) that potentially can influence a user's decision to adopt a technology or not. This brought about the formulation of the Unified Theory of Acceptance and Use of Technology (UTAUT). UTAUT aims to present the mediators and factors that impact a user's perception of the adoption of a given technology, and its usage or eventually act as a determinant in its acceptance. Alatawi et al. (2012) noted that UTAUT provides an integrated theoretical base to advance research that is premised on information and communication technology (ICT) diffusion and adoption. This takes into account the weaknesses and lapses associated with other models or theories for technology acceptance, adoption, and use. UTAUT is an embodiment of eight theories for technology acceptance and behaviour. This consolidated these theories based on empirical and conceptual viewpoints. UTAUT is made up of the following models/theories:

- Theory of Planned Behaviour (TPB)
- Theory of Reasoned Action (TRA)
- Motivational Model (MM)
- Technology Acceptance Model (TAM)
- Combined TAM and TPB (C-TAM-TPB)
- Innovation Diffusion Theory (IDT)
- Model of PC Utilisation (MPCU)
- Social Cognitive Theory (SCT) (Venkatesh et al., 2003).

The theory considered the impact of mediating factors, most of which did not attract attention in the previous theories/models on technology adoption but proved to be important. Accordingly, Jong (2009) stated that among a wide range of technology adoption theories, UTAUT appears to be the best fit, as it entails detailed knowledge of technology adoption. In the development of UTAUT, four major constructs served as the basis of the theory: (1) performance expectancy; (2) effort expectancy; (3) social influence; and (4) facilitating conditions. Also, the significant moderators embedded in the model are age, gender, experience, and voluntariness (Figure 5.6). A close examination of the theory shows a detailed and comprehensive perspective of the evolution of the determinants of behaviour and intentions over time. It is pertinent to emphasise that a high number of the germane relationships making up the model are moderated.

An example is the inclusion of age, which received little consideration in the previous technology adoption models. However, the outcome of the study on UTAUT is that it brings about moderation in all the important relationships in the model. Likewise, the inclusion of gender was not initially considered in previous theories has served as a key moderating factor.

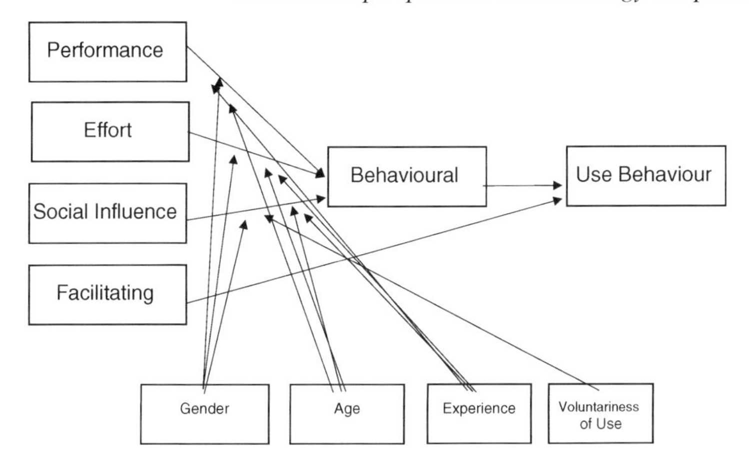

Figure 5.6 Theory of Acceptance and Use of Technology (UTAUT)
Source: Venkatesh et al. (2003).

Technology adoption studies in architecture, engineering and construction (AEC) industries

A couple of research studies have been carried out on technological acceptance or adoption in the architecture, engineering, and construction (AEC) industries. In recent times, a wide range of these studies have focused on the adoption of building information modelling (BIM), since this appears to be the rave of the moment in construction circles. These research studies identified several factors that influence an individual's perception and expected outcome of technology use. These studies identified performance expectations, performance efforts, facilitating conditions, and social influence as having significant roles in determining the adoption and usage of ICT in the AEC industry.

Son et al. (2012) applied TAM in the examination of user satisfaction of South Korean construction professionals in the use of portable computing devices. The findings from the study revealed that "ease of use" and "usefulness" both are related to user satisfaction. Also, the study indicated that a more predictive behaviour is perceived usefulness compared to ease of use. Lowry (2012) used Innovation Diffusion Theory (IDT) to discover the acceptance level of building management systems by users. Using a questionnaire, the study used the perceived attributes of BIM from IDT and showed that predicted usage is not the result of an individual's belief. According to the study, voluntariness proved to be the lone predictive variable that refutes the prediction of TAM, noting that the use of building management systems by users is only done when it is explicitly required. Davis and Songer (2008) proposed the "social architecture" factor model based on a rigorous review of TAM in conjunction with TRA and TPB. The proposed model explained both individual and organisational dispositions towards adoption of technology or resistance to technology. The study used a questionnaire survey made up of 554 items to measure individual differences with respect to beliefs and attitudes towards ICT in identifying users' resistance to the adoption of ICT.

Jacobsson and Linderoth (2012) ascertained the satisfaction level of users in Sweden with ICT, using surveys and interviews. The study revealed that users were generally satisfied with the inclusion of ICT in their work, although the use of ICT brought about an increase in competitiveness. Equally, these authors (ibid.) researched the adoption of ICT in construction projects in Sweden. Emphasis was put on the importance of personal evaluation of technology as a defining factor for its development and use. One very important finding of the study is the prominence of project-defined reference that gave prominence to action and time savings, which ultimately influenced the adoption and implementation. Peansupap and Walker (2005) used two survey questionnaires. The first entailed 46 items spanning various IDT variables and included self-motivation, measurement of external variables, and variables making up knowledge management approaches. The second entailed examining the perceived value attached to construction project information management, revealed through surveying large-scale construction outfit participants and project management outfits to rate the significance of variables making up a particular construct. The study results indicated that the significance of TAM studies matches users' belief in the importance of acceptance factors. It also revealed that the benefits of an adopted technology are ranked higher than user support or the technology characteristics.

Samuelson and Björk (2013) reviewed the adoption of electronic data exchange, electronic document management, and BIM. The study interviewed technology managers in 11 construction organisations in Sweden. One interesting finding from the study is that BIM adoption is initially "bottom-up", aimed at attaining an individual's professional benefit, after that, the organisations want to benefit from the initiative. Walter (2006) investigated the adoption of ICT using innovation theory in three construction companies using a case study. The emphasis of the study was placed on the implementation and the strategic management of ICT rather than the acceptance by an individual. Adopting TAM, the study scrutinised users' adoption that was the result of the decision taken by organisations to adopt ICT. Also, the prominence of social networks and their influence on the adoption choice were highlighted by the study.

A study conducted in Australia by Brewer and Gajendram (2011) used an interview-based case study to identify impediments to engaging individual users:

> Despite the presence of various state-of-the art ICT systems including BIM, the project participants were largely ignorant of their presence and would have been unlikely to use them if they had known of their existence, due to a lack of conviction as to their benefit.
>
> (ibid.)

The study revealed that personal features, beliefs, and values impacted the decision to adopt BIM. A model was proposed by the study, which centred on adoption and use, connection at the group level was established and stressing the organisational culture of the project, of which a constituent part is the individual's expectations and beliefs. One critical factor identified in cultural analysis is social influence.

Miller et al. (2009) investigated the implementation of four computer-aided design (CAD) technologies in large construction organisations in Australia, using a case study. The study identifies "usability", "benefit". and "value" as the main constructs that serve as determinants of a successful implementation. The study proposed a model of adoption termed "perception influence" that recommends a series of layers that consists of usability, personal benefit, project benefit, and organisational value. The study found the successful implementation of one technology. In contrast, the others failed due to being rejected by the project staff or were stopped because of the negative disposition of the benefits accrued from the management of the project. Lee et al. (2015) proposed a BIM Acceptance Model (BAM) for the South Korean construction industry. The model highlighted 28 main factors focusing on TAM external variables. The factors are: top management support, output quality, internal pressure, external pressure, personal innovativeness, collective efficiency, self-efficacy, collective efficiency, and compatibility. The study's conclusion stated that perceived usefulness has a potent impact on a person's intention to adopt BIM. Furthermore, the study revealed that organisational competency proved to be highly significant with regards to a person's perceived ease of use, which translates as, the better the organisation implements BIM, the easier it is to use it.

Summary

The chapter gave a detailed description of the theories and models on technology adoption propounded over time. Furthermore, chronological and sequential details of these theories and models, based on how they have evolved over time, were given. The reviewed theories and models include: Innovation Diffusion Theory, Social Cognitive Theory, the Theory of Reasoned Action, the Theory of Planned Behaviour, the Task Technology Fit Model, the two Technology Acceptance models, and the Unified Theory of Acceptance and Use of Technology. Based on the review of the theories and models, the UTAUT was adopted as the underpinning theoretical basis for the current study.

References

Ajzen, I. (1985). From intentions to actions: A theory of planned behavior. In J. Kuhl and J. Beckmann (eds), *Action Control*. Berlin: Springer, pp. 11–39.

Ajzen, I. (1991). The theory of planned behavior. *Organisational Behavior and Human Decision Processes*, 50(2), 179–211.

Ajzen, I. (2002). Perceived behavioral control, self-efficacy, locus of control, and the theory of planned behavior. *Journal of Applied Social Psychology*, 32(4), 665–683.

Ajzen, I. and Fishbein, M. (1980). *Understanding Attitudes and Predicting Social Behavior*. Englewood Cliffs, NJ: Prentice-Hall.

Alatawi, F.M.H., Dwivedi, Y.K., Williams, M.D., and Rana, N.P. (2012). *Conceptual model for examining knowledge management system (KMS) adoption in public sector organisations in Saudi Arabia*. Paper presented at the tGOV Workshop '12 (tGOV12), Brunei University, West London.

Alhassany, H. and Faisal, F. (2018). Factors influencing the internet banking adoption decision in North Cyprus: Evidence from the partial least square approach of the structural equation modeling. *Financial Innovation*, 4(9).

Autry, C.W., Grawe, S.J., Daugherty, P.J., and Richey, R.G. (2010). The effects of technological turbulence and breadth on supply chain technology acceptance and adoption. *Journal of Operations Management*, 28(6), 522–536.

Bandura, A. (1986). *Social Foundations of Thought and Action: A Social Cognitive Theory*. Englewood Cliffs, NJ: Prentice-Hall, Inc.

Brewer, G. and Gajendram, T. (2011). Attitudinal, behavioural, and cultural impacts on e-business use in a project team: a case study. *Journal of Information Technology in Construction (ITcon)*, 16(37), 637–652.

Chan, S., and Lu, M. (2004). Understanding internet banking adoption and use behavior: A Hong Kong perspective. *Journal of Global Information and Management*, 12, 21–43.

Chau, P. and Hu, P. (2002). Examining a model of information technology acceptance by individual professionals: An exploratory study. *Journal of Management Information Systems*, 18(4),191–229.

Davis, F., Bagozzi, R., and Warshaw, P. (1989). User acceptance of computer technology: A comparison of two theoretical models. *Journal of Management Science*, 35(8), 982–1003.

Davis, K.A. and Songer, A.D. (2008). Resistance to it change in the AEC industry: An individual assessment tool. *Electronic Journal of Information Technology in Construction*, 13, 56–68.

Eben, A.K. and Achampong, A.K. (2010). Modelling computer usage intentions of tertiary students in a developing country through the technology acceptance model. *International Journal of Education and Development Using Information and Communication Technology*, 6(1), 102–116.

Goodhue, D. and Thompson, R. (1995). Task-technology fit and individual performance. *MIS Quarterly*, 19(2), 213–236.

Ikuabe, M., Aigbavboa, C., Anumba, C., Oke, A. and Aghimein, L. (2022). Confirmatory factor analysis of performance measurement indicators determining the uptake of CPS for facilities management. *Buildings*, 12(4), 466.

Jacobsson, M. and Linderoth, H.C.J. (2012). User perceptions of ICT impacts in Swedish construction companies. *Construction Management and Economics*, 30(5), 339–357.

Jong, D. (2009). *The acceptance and use of the learning management system*. Paper presented at Fourth International Conference on Innovative Computing, Information and Control, Kaohsiung, Taiwan, 7–9 December 2009.

Kamara, J., Anumba, C., and Evbuomwan, N. (2001). Assessing the suitability of current briefing practices in construction with a concurrent engineering framework. *International Journal of Project Management*, 19(6), 337–351.

Kamara, J., Heidrich, O., Tafaro, V., Maltese, S., Dejaco, M., and Cecconi, F. (2020), Change factors and the adaptability of buildings. *Sustainability*, 12(6), 6585. http://dx.doi.org/10.3390/su12166585

King, W.R. and He, J. (2006). A meta-analysis of the technology acceptance model. *Information & Management*, 43(6), 740–755.

Leach, S., Stewart, J., and Walsh, K. (1994). *The Changing Organisation and Management of Local Government*. London: Macmillan.

Lee, J., Jeong, Y., Oh, Y., Lee, J., Ahn, N., Lee, J., and Yoon, S. (2013). An integrated approach to intelligent urban facilities management for real-time emergency response. *Automation in Construction*, 30, 256–264.

Lee, S., Yu, J., and Jeong, D. (2015). BIM acceptance model in construction organisations. *Journal of Management in Engineering*, 31(3), 4014048.

Lee, Y.H., Hsieh, Y.C., and Ma, C.Y. (2011). A model of organisational employees' e-learning systems acceptance. *Knowledge-Based Systems*, 24(3), 355–366.

Legris, P., Ingham, J., and Collerette, P. (2003). Why do people use information technology? A critical review of the technology acceptance model. *Information and Management*, 40(2), 191–204.

Liu, Z., Min, Q., and Ji, S. (2008). *A comprehensive review of research in IT adoption*. Proceedings of the 4th International Conference on Wireless Communications, Networking and Mobile Computing, pp. 1–5.

Lowry, G. (2012). Modelling user acceptance of building management systems. *Automation in Construction*, 11(6), 695–705.

Magee, A. (2002). Attitude-behaviour relationship. Available at: http://www.ciadvertising.org/SA/fall_02/adv382j/mageeac/theory.htm.

Mathieson, K. (1991). Predicting user intentions: Comparing the technology acceptance model with the theory of planned behavior. *Information Systems Research*, 2 (3), 173–191.

Miller, A., Radcliffe, D., and Isokangas, E. (2009). A perception-influence model for the management of technology implementation in construction. *Construction Innovation*, 9(2), 168–183.

Moore, T. (2012). Towards an integrated model of IT acceptance in healthcare. *Decision Support Systems*, 53, 507–516.

Peansupap, V. and Walker, D.H. (2005). Factors enabling information and communication technology diffusion and actual implementation in construction organisations. *Journal of Information Technology in Construction (ITcon)*, 10(14), 193–218.

Rogers, E.M. (1995). *Diffusion of Innovation*. 4th edn. New York: The Free Press.

Rogers, E.M. (2003). *Diffusion of Innovation*. 5th edn. New York: The Free Press.

Saeed, K.A. and Abdinnour-Helm, S. (2008). Examining the effects of information system characteristics and perceived usefulness on post adoption usage of information systems. *Information and Management*, 45(6), 376–386.

Samuelson, O. and Björk, B.C. (2013). Adoption processes for EDM, EDI and BIM technologies in the construction industry. *Journal of Civil Engineering and Management*, 19(1), 172–187.

Son, H., Park, Y., Kim, C., and Chou, J.S. (2012). Toward an understanding of construction professionals' acceptance of mobile computing devices in South Korea: An extension of the Technology Acceptance Model. *Automation in Construction*, 28, 82–90.

Venkatesh, V. and Davis, F. (2000). A theoretical extension of the Technology Acceptance Model: Four longitudinal field studies. *Management Science*, 46(2), 186–204.

Venkatesh, V., Morris, M.G., Davis, G.B., and Davis, F.D. (2003). User acceptance of information technology: Toward a unified view. *MIS Quarterly*, 27(3), 425–478.

Wallace, L.G. and Sheetz, S.D. (2014). The adoption of software measures: A technology acceptance model (tam) perspective, *Information & Management*, 51(2), 249–259.

Walter, M. (2006). Return on interoperability: The new ROI. *CAD USER*, 19(3), 14.

Xu, H., Feng, J., and Li, S. (2014). Users-orientated evaluation of building information model in the Chinese construction industry. *Automation in Construction*, 39, 32–46.

Yzer, M. (2012). Perceived behavioral control in reasoned action theory: A dual-aspect interpretation. *The Annals of the American Academy of Political and Social Sciences*, 640(1), 101–117.

6 Gaps in technology adoption research

Introduction

With particular emphasis on facilities management (FM), after a comprehensive review of theories and models on technology adoption, there appears to be something missing regarding the prevailing peculiarities attributed to the management of built infrastructures. The use of a particular technology in meeting the objectives set out by an organisation is overwhelmingly influenced by the core tenets and principles guiding FM. Similarly, the post-occupancy stage of the building life-cycle has its own peculiar and fundamental attributes, which make it distinct when discussing adopting a particular technology. According to Ashworth and Tucker (2017) who emphasised the adoption of building information modelling (BIM) for FM; some important tenets of FM were not duly given painstaking consideration for the technological innovation which has long been used in certain aspects of the management of facilities. Studies on technology adoption for different construction phases are not particularly new; several studies have been carried out in this field. Several postulated theories such as the Theory of Planned Behaviour (Taylor and Todd, 1995), the Theory of Reasoned Action (Taylor and Todd, 1995), Social Cognitive Theory (Bandura, 1989), Innovation Diffusion Theory (Rogers, 1995) and the Technology Adoption Model (Davis et al., 1989) all centre on the explanation and prediction of the behaviour to adopt information technology, whereas the Unified Theory of Acceptance and Use of Technology (UTAUT) is an integrated model formulated and tested empirically to explain the adoption of information technologies (Venkatesh et al., 2003). The explanation provided by the UTAUT on technology adoption appears to be more concise and robust than that of the other individual theories or models (Venkatesh et al., 2012). However, these different theories and models for technology adoption suffer from inconsistency and uncertainty in the prediction and explanation of the adoption of certain technologies in different contexts (Turner et al., 2010). These theories or models are based on an approach that is variance-defined (Beaudry and Pinsonneault, 2005), basically outlining the core determinants and establishing their relationship with regard to behaviour for the adoption of technology.

DOI: 10.1201/9781003376262-10

Furthermore, these theories emphasise the assumption of positive eventualities of the technology adoption (Sanakulov and Karjaluoto, 2015), and do not consider the effects of experimental factors on the behaviour concerning technology adoption. According to Parijat and Bagga (2014), experimental factors can influence certain behaviours with regards to job motivation and the behaviour of a consumer. Also, the group of theories and models dwell more on the adoption of technology at the system level. It is assumed that behavioural disposition at the system level is relatively simple compared to the feature level which has more forms and dynamism. At the feature level, variants of form for behaviour for adoption can include the trial of a new feature, the combination of both new and old features, and the substitution of an old feature with that of a new feature to accomplish a particular task. With this in mind, this book seeks to introduce two new dimensions into the discussion of technology adoption, with an emphasis on cyber-physical systems (CPS) for FM. These constructs are the business environment factors and performance measurement indicators of the adopted technology.

Gap one: the business environment

There is growing necessity to foster awareness of the influential function of facilities at all levels of the management of organisations. Du et al. (2010) noted that when attempting to attain competitive advantage, organisations put in tremendous efforts to optimise business resources. In contrast, the marketplace dynamics, coupled with the fast pace of technological advances, force many organisations to take a better stance on their facilities and operational assets. There is fast becoming a consensus among the top executives of organisations that deploying operational facilities as a business resource is the best approach to raise awareness of corporate management, in the same way as technology and human resources in business resources (ibid.). Thus, the importance of the acknowledgement by top executives that facilities denote a business concern awaiting use has been established. To improve the competitiveness of organisations, key business restructuring and philosophy for the improvement of effectiveness have been introduced. It is quite clear that in most organisations, facilities are no longer of negligible significance, even though the focus is usually on the main activities. There strategic function of FM is recognised, while there is a better understanding of the provision of opportunities through effective management.

According to Turpen et al. (2016), management must acknowledge that facilities form a core component of the business resources. Facilities managers need to have controlled powers on strategic decisions and also make contributions that would help facilities meet business targets and organisational objectives. In response to the ever-changing business applications, the scope and range of facility functions should be extended beyond the provision of technical solutions to issues arising and should ensure that facilities' effectiveness is maximised and the cost of occupancy is minimised. A vital aspect of operational assets that calls for consideration as a strategic resource is their financial performance impacts on the organisation, based on the asset use (Schaltegger and Roger, 2017). One of the roles of FM is to tackle organisations' business challenges. On a long-term basis, the role of sustainable FM in organisations should be based on drive and pursuit in adding value, by making provisions for innovative and appropriate "facilities solutions" to challenges confronting business, by skilful manipulation of resources at the disposal of the business – maintaining ideal stability between people, assets. and technology.

The business environment is characterised by uncertainties, which has led organisations to be intentional about and better informed of the effects of the environmental forces that are prevalent in corporate life and their attendant influence on the attainment of projected goals. Samsami et al. (2015) noted that there is variance in environmental ambiguities across several sectors. The alternating specification is a contributory factor in determining how simple or complex are the scenarios encountered by the organisation. Putting the discussion of environmental factors in a simple perspective, they can be categorised into several forms on the basis of their outlined criteria. Environmental factors denote the factors associated with the organisation's environment, which refers to the delineated external effects and situations that affect the life and development of the organisation (Kyrgidou and Spyropoulou, 2013). Some of the important factors are economic, political, legal, and technological environments. Moreover, the micro environmental factors are concerned with the characteristics of several sectors. These factors make up the internal conditions and effects experienced in the development of the sector. It is argued that the economic consideration is the most vital factor among these factors (Barrett and Dannenberg, 2017).

Göpel (2016) noted that efforts are made by organisations to affirm conformance of strategies that are in alignment with the *ad infinitum* altering environment. Due to resources not being readily available in the event of an economic downturn, management's fast reaction to the realities leads to assurances of better delivery and efficiency through the provisions given by emerging opportunities. Organisations' long-term survival and their position of competitive advantage are endangered when management reneges on evaluating the delivery of the functional activities of the organisation, such as marketing, finance, and production (Tapera, 2014). Therefore, the syncretisation of the business environment in conjunction with corporate strategies in delivering outlined goals on a short- and long-term basis is important to the organisation (Kabeyi, 2019). The combination of the organisation's resources and its business environment is known as a strategy (Islami et al., 2020). However, identifying the various environmental influences can be an enormous task, due to the preference of managers to create an image based on the organisation's vital influences. The organisation's operations are conducted within the confines of their business environment, and the relevance of the various influences might change over time. The creation of possible scenarios that influence the strategies to be used by organisations to adapt to a new reality is carried out after the assessment of the relevant environmental factors (Samsami et al., 2015).

Contingency theory

The business environment is dynamic in nature, which has compelled organisations to constantly adapt to the ensuing circumstances (Teece, 2018). These dynamic forces and global competitiveness have led to the need for a rerouting in implemented strategies to help organisations take the appropriate measures and use proper resources, permitting quick responses to changes in the marketplace, the production of high-quality goods, and innovative designs (Paiva et al., 2008). The organisation's pursuit of success in areas characterised by fierce competition must be based on operating strategies that align with the operations of the external environment, and also with positioning of resources to support it (Awwad et al., 2013). Ireland and Hitt (2005) affirmed that there is a strong relationship between strategy elements and the environment externalities in the management of the organisation. Also, the need to align the attributes of the environment and the organisation's operations has been constantly proved to be a significant principle of management thought (Barrett and Dannenberg, 2017). In the context of contingency theory, the operations of an organisational structure are seen from the perspective of the organisation's strategy, the environment, and size, while they are referred to as factors that organisations must adapt to.

Organisations' adaption to the environment is seen at two levels. The first is the sub-units that make up the organisations to match the environmental attributes, while the second is the structural formation of the organisation in conformance to the general environment where its operations are conducted (Lengnick-Hall et al., 2020). Using contingency theory as a basis, an organisation's conformity to contingencies leads to enhanced and better delivery, therefore, "fit" is achieved and dissonance with the environment is avoided (Donaldson, 2001). Therefore, organisations are shaped by their environment since they are compelled to adapt to the environment to deliver the best performance. Moreover, avoiding loss of performance stimulates organisations to evolve, leading to changes in the business environment (Lamberg et al., 2021). The allocation of resources is also important to the organisation due to the substantial implications for the growth or failure of the business in the choice of resources. This is mostly experienced in organisations where investment in a particular service or product subsequently experience a downturn, due to the disappearance of the market in a hostile business environment. This occurs when the organisation undertakes strategic management activities that entail the allocation of major irreversible resources (Kabeyi, 2019).

Contingency theory propounds that the strategy of organisations is influenced by the business environment (Iqbal et al., 2015). According to Bess and Dee (2008, p. 138), contingency theory is "a process of achieving a 'fit' between the continuous management of the environment and the organisation's design". The theory assumes that different solutions might be effective (Dobak and Antal, 2010). This is regarded as one of the fundamental principles of the theory because instead of affirming the application of universal principles of organisation management, it affirms that different organisational structures should be required in different circumstances (Cosh et al., 2012). Furthermore, an essential idea of contingency theory is that the viability of organisations depends on the appropriate match between the environment and the organisation. Organisations are deemed to be an open system, which "stresses the complexity and variability of the individual parts – both individual participants and subgroups – as well as the looseness of the connections among them" (Scott, 2003, p. 101). The viability of any organisation depends on its ability to visualise and ensure the incorporation of the "contingencies" prevailing in the environment. Moreover, in attaining success in a dynamic and rapidly evolving environment, organisations must be internally dynamic and flexible, coupled with capabilities in innovations and renewals. However, various organisations are premised in different markets and various management styles as well as having different compositions of staff members. Hence, to follow contingency theory, organisations should monitor the environment and realise that the handling of different situations must be attended to in different ways.

Competitive advantage theory

Organisations can attain a competitive advantage by synergising their adopted strategies with the organisation's internal and/or external business environment. Hence, achieving competitive advantage can be viewed from different standpoints. From the perspective of market position, competitive advantage has a basis in industrial organisation economics, where market positions or mobility barriers are important sources of competitive advantages that enhance delivery and performance (Sigalas and Economou, 2013). This predicts that competitive advantage is attributed to external attributes or characteristics. The acclimatisation of organisations to industry or market pressures and demands ensures their survival and growth. Meanwhile, those that fail to adapt are bound to fail and witness exit from the industry or market (Lipczynski et al., 2005). It is argued that organisations can achieve competitive advantage by offering a set of distinct services or products or products at low cost, or by providing services to a particular group of customers. In this instance, competitive advantage is more attributed to external features than to the idiosyncratic competencies of the organisation and its resource-based dispositions. Furthermore, another aspect focuses on the competency and resources of the organisation as a basis for its competitive advantage. However, the approaches mentioned above are complementary, with the former giving insight into the value of competitive results in the business environment. At the same time, the latter outlines the dynamic aspects of the organisation's attitude concerning the accrual and nature of the organisation's resources (Chen et al., 2021). A typical example is the use of resources in various products and markets, coupled with the fact that products and markets might need different resources.

Technological innovations may lead to environmental instability by accelerating the rate of change in scientific communities in a business environment. Technological diffusion has undergone an intensified rate of change with particular emphasis on an information-intensive sector (Hitt et al., 2009). This highly competitive business landscape instability leads to uncertainty and high risk during the process of strategic planning, hence, strengthening the call for high-level scanning of the environment coupled with a more proactive approach (Lindelof and Lofsten, 2006). The attainment of sustainable competitive advantage hugely depends on the ability of the firm to adapt to the changing environment. This reinforces the idea of a managerial attitude that is focused on innovativeness (Kuratko and Audretsch, 2009). This can be actualised by organisations turning to entrepreneurial orientation. Giovannetti et al. (2022) emphasised process innovation and highlighted drivers for process innovation include customer-driven process change, process time reduction, a restructuring that is process-based and financially driven, process cost reduction, process redesign prior to initiating outsourcing activities, lean production as a tool for process redesign used for cross-functional solutions, and process change before the implementation of software systems. On this basis, there is a clear distinction between process innovation and process improvement. Process improvement stirs up change at a lower scale and usually starts with the existing process, while process innovation reflects business process reengineering, it brings about a thorough change even without an existing process as the starting point. Innovation and the different strategies used in implementation come in phases of competitive challenges.

Wales et al. (2020) stated that being proactive displays an organisation's aggressiveness in the pursuit of sustainable competitiveness. Being proactive is seen as an organisation's stance in anticipation and preparation and acts on future wants and demands from the market and the creation of first-mover advantage (Fatoki, 2014). Moreover, being proactive is an action direction related to competitive superiority arising from the tactics used to be "one step ahead" and the exhibited market leadership attributes by organisations showing these forms of strategic behaviour (Vinayan et al., 2012). Proactive organisations recognise potential and current clients' future needs, anticipate changes in demand, and engage in trend monitoring. There is a powerful relationship between strategic management and the dimensions of a proactive stance of an organisation's orientation (Osita-Ejikeme, 2021). A distinct attribute of managers who are proactive is that they have their eye on future opportunities to exploit performance improvement and growth, thereby creating competitive advantage (Teece, 2018). Being proactive leads to competitive advantage by forcing competitors into responding to initiatives by the first mover. The first-mover advantage is regarded as the benefits accrued by organisations that are the first to engage in the production of a new service or product, the adoption of new technologies for operations, entrance into a new market, or establishing a brand identity (Tetrault Sirsly and Lamertz, 2008).

Gap two: performance measurement

The performance of an organisation is hugely dependent on its ability to gather and integrate a high flow of management information structured for the intelligent actions based on the information. Performance is the process or action to bring to fruition or deliver on a function or task (Marques-Quinteiro et al., 2013). Moreover, task performance is the ability to efficiently deliver functions following the outlined task description, while also recognising the use of required knowledge and skills (Hubka and Eder, 2012). Also, Armstrong and Murlis (2000, p. 240) noted that the conclusion of a task "with predetermined procedures" denotes performance. While Ismajli et al. (2015) noted that elaborate support of performance is a significant variable in skills, knowledge, and motivation. Motivation in this regard refers to the commitment to deliver on an obligation, whereas skills denote the ability to engage in a function, while knowledge is the possession of the requisite details pertaining to what should be done. In general, the overview of performance is the integration of these three factors (Dubnick, 2005).

Aguinis (2019) noted that performance measurement is carried out as a means to maintain and monitor the organisation's control, which translates to the process of ensuring the successful attainment of its overall objectives and goals through organisational strategies. With the view of a change in organisations' focus, Tuan (2010) posited that an important agent of change is performance measurement. After focus has been attained by an organisation, performance measurement plays a pertinent role in giving stability and maintaining the focus on changing customer demands and requirements and actions taken by competitors. Furthermore, Muchri et al. (2011) observed that performance measurement is a vital factor in ascertaining the efficacious execution of an organisation's strategy. The performance of business units should be assessed, based on the outlined objectives formulated during planning. Bourne et al. (2003) noted that inappropriate performance measurement serves as a barrier to the organisation's development since measurement serves as a link between actions and strategies. Inappropriate measurement brings about actions that do not support the adopted strategies, even though they were articulately formulated and communicated. The adoption of appropriate measures ought to provide and strengthen links, which brings about the achievement of strategic objectives and the impact on the strategies and goals needed for their achievement. Gunasekaran et al. (2004) advocate that the role of measurement is vital in the improvement of productivity and quality to provide standards to enable comparisons; ensure the attainment of customer needs; provide a "scoreboard" to monitor performance; outline problems and determine areas that need urgent attention. This scoreboard gives an indication of the costs associated with poor quality; justifies the use of resources; and provides feedback for the promotion of effort improvement. Furthermore, Park et al. (1996) observed that the reasons for the evaluation of performance are:

- To characterise and understand products, processes, the environment, and resources, and also to establish a basis for comparing future assessments.
- To evaluate and determine status in line with a stipulated plan.
- To predict the enablement of planning.
- To improve support by collating information which aids problem identification and also the planning and tracking of efforts improvements.

In all these classifications, Neely (1998) noted that the highlighted reasons are based on one of these outlined generic classifications:

- to compel progress;
- to confirm priorities;
- to communicate position;
- to check position.

The need for information to inspire suitable action and organisational learning at the appropriate level of the organisation and the stage of making decisions underscores the importance of performance measurement (Aguinis, 2019). In a business environment that is competitive, management-by-exception might be put in place as against an interactive systems.

The Balanced Scorecard

There is a common argument that performance measurement ought to be an offshoot of strategy, meaning that they should be used to reinforce the relevance of some strategic objectives (Talbot, 2010). According to Kaplan and Norton (1996), the Balanced Scorecard shows a model for strategic performance measurement and management in high performance organisations. It translates the vision of the organisation into a group of performance indicators which are divided into four different perspectives (Figure 6.1):

- How customers are to be viewed (Customers).
- The internal processes to ensure success will be achieved (Internal Processes).
- How the shareholders will be viewed (Financial).
- Methods and strategies to improve the organisation (Innovation).

By assessing an organisation using the outlined perspectives, the Balanced Scorecard aims to connect short-term operation control with the long-term strategy and vision of the organisation.

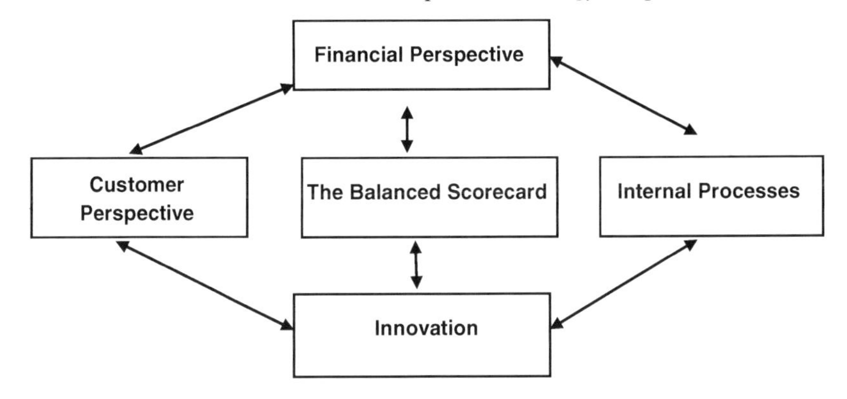

Figure 6.1 Perspectives of the Balanced Scorecard
Source: Kaplan and Norton (1996).

The Balanced Scorecard translates the strategy and mission of an organisation into a set of comprehensive performance measures and provides a framework of strategic measurement and management. According to Bhagwat and Sharma (2007), the Balanced Scorecard provides:

- a practical guide for the implementation of corporate strategy;
- a management tool to connect the general objectives for individuals, teams, and business, and also the rewards for strategic goals;
- an efficient system to implement change management;
- better understanding of the factors driving business success;
- easy identification of "cause-and-effect" interdependencies in operations;
- qualitative and quantitative information;
- dynamism in communication and response.

The Balanced Scorecard serves as a management system that measures an organisation's operations and economic performance. It strives to make a compelling argument for the addition of measures of a non-financial basis in the overall measurement system of an organisation. The system's strength arises from a second "balance" that supersedes the ad-hoc measures of financial and non-financial terms. This system portrays the intricacies of the organisation's strategy, which is conveyed through a model of cause-and-effect, which connects the measures to the value of the shareholder. Measures that are non-financial, such as employee turnover and customer retention, are attributed to the scorecard to reflect activities performed by an organisation in executing its strategy. Thus, the measures act as predictors for future performance on a financial basis.

The "SMART" pyramid

SMART means *S*trategic *M*easurement *A*nalysis and *R*eporting *T*echnique and it was developed by Wang Laboratories (Ghalayini and Noble, 1996).

This was pursued due to overwhelming dissatisfaction with traditional performance measurements, which include productivity, efficiency, utilisation, and certain financial variances. The aim was to formulate a management control system exhibiting performance indicators premeditated for the definition and sustenance of success (Taticchi and Balachandran, 2008). The SMART system represents a four-layered pyramid of measures and objectives. The system follows the idea of taking measurements from the bottom level of an organisation, at the centre, work and department levels, which reflects the corporate mission coupled with external and internal business objectives (Taticchi and Balachandran, 2008).

The apex of the pyramid is the strategy or vision of the organisation. This layer involves the assignment of roles of a corporate portfolio by management to individual business units and the allocation of resources to support them. The second layer defines the objectives of an individual business unit concerning finances and markets. The third layer is the definition of more tangible operating priorities and objectives concerning individual business operations systems based on productivity, flexibility, and customer satisfaction. The fourth layer denotes productivity, flexibility, customer satisfaction, and departmental level by criteria of specific operations: cost, time, quality, and delivery. The fifth layer is the foundation of the pyramid, which exhibits the operational measures; these are vital for achieving high-level results and ensuring the implementation success of the strategy adopted by the organisation.

The performance prism

According to Moura (Moura et al., 2019), the future attainment of competitive success is dependent on embarking on an inclusive approach in management, reflecting the need to consider the requirements of all stakeholders to be part of management functions and performance measurement. The performance prism includes the perspective of stakeholders' involvement in performance measurement to reflect the importance of meeting stakeholders' satisfaction (O'Boyle and Hassan, 2013). After identifying the organisation's key stakeholders and their outlined requirements, it is important to consider the likelihood of the organisation's strategies of delivering satisfaction to the stakeholders. The need to implement the measures engaged in the reflection and communication of the organisation's strategies is a consistent topic when discussing performance management. Kennerley and Neely (2000) outlined five different but connected performance perspectives that organisations need to consider when defining a set of performance measures:

- *Stakeholders satisfaction* – Defining who the key stakeholders are and what they need or want.
- *Strategies* – Outlining the strategies needed to be put in place to achieve the satisfaction of the key stakeholders.

- *Processes* – Identifying the critical processes needed for the operation and enhancement of these processes.
- *Capabilities* – Engagement of capabilities required for the operation and enhancement of these processes.
- *Stakeholder contribution* – Identifying the contributions required from the stakeholders for the maintenance and development of these capabilities.

The EFQM model

The European Foundation for Quality Management (EFQM) model assesses performance from a broader perspective by analysing areas of performance not considered in methodologies. The model defines the enabling factors of performance improvement and identifies result areas that ought to be measured, as shown in Figure 6.2. The EFQM model has its basis on nine individual criteria; five of these serve as enablers (ways of doing things in the organisation) while four serve as a result (achievement by enablers). The idea is centred on "excellent results concerning performance, customers, people, partnerships and resources, and processes" (EFQM, 1999). Furthermore, the enablers are described as processes, resources, and policies, which transform inputs into outputs, and the organisation's outcome and output are indicated as the measure of the result (Heras-Saizarbitoria et al., 2012). The process is methodical, having the individual nine criteria matched up with sub-elements to continuously indicate the level of attainment. The assessment process is unambiguous in the requirement of services for continuous improvement, evidence of quality issues, or trend data.

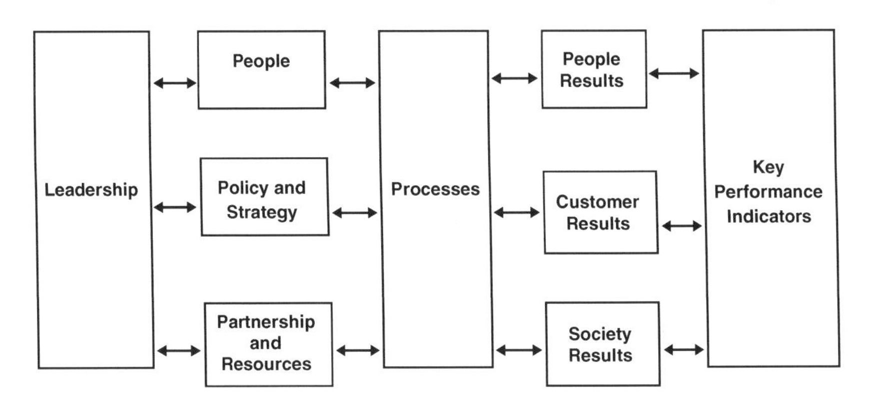

Figure 6.2 The EFQM model
Source: (EFQM, 1999).

Summary

This chapter explored the identified gaps in the literature on technology adoption. This was premised on revealing the gaps peculiar to adopting cyber-physical systems for facilities management. Business environment and performance measurement were revealed as key to technological adoption. Moreover, considering the peculiar nature of facilities management as a multidisciplinary and multifaceted function whose importance to actualising the organisation's objectives cannot be over-emphasised, the features of the business environment and performance measurement are vital in the discussion of technology adoption. Hence, the resulting findings of the chapter serve as a basis for formulating the hypothesised model for the book.

References

Aguinis, H. (2019). *Performance Management for Dummies.* Chichester: John Wiley & Sons.

Armstrong, M. and Murlis, H. (2000). *Reward Management: A Handbook of Remuneration Strategy and Practice.* London: Hay Group.

Ashworth, S. and Tucker, M. (2017). *FM Awareness of Building Information Modelling (BIM).* 1st edn. BIFM.

Awwad, A., Al Khattab, A., and Anchor, J. (2013). Competitive priorities and competitive advantage in Jordanian manufacturing. *Journal of Service and Management*, 6(1), 69–79.

Bandura, A. (1989). Human agency in social cognitive theory. *American Psychologist*, 44(9), 1175–1184.

Barrett, S. and Dannenberg, A. (2017). Tipping versus cooperating to supply a public good. *Journal of the European Economic Association*, 15(4), 910–941.

Beaudry, A. and Pinsonneault, A. (2005). The other side of acceptance: Studying the direct and indirect effects of emotions on information technology use. *MIS Quarterly*, 34(4), 689–710.

Bess, J.L. and Dee, J.R. (2008). *Understanding College and University Organisation: Theories for Effective Policy and Practice.* Virginia: Stylus Publishing, LLC.

Bhagwat, R. and Sharma, M. (2007). Performance measurement of supply chain management: A balanced scorecard approach. *Computers & Industrial Engineering*, 53(1), 43–62.

Bourne, M., Neely, A., Mills, J. and Platts, K. (2003). Implementing performance measurement systems: a literature review. *International Journal of Business Performance Management*, 5(1), 1–24.

Chen, M., Michel, J., and Lin, W. (2021). Worlds apart? Connecting competitive dynamics and the resource-based view of the firm. *Journal of Management*, 47(7), 1820–1840.

Cosh, A., Fu, X., and Hughes, A. (2012). Organisation structure and innovation performance in different environments. *Small Business Economics*, 39(2), 301–317.

Davis, F., Bagozzi, R., and Warshaw, P. (1989). User acceptance of computer technology: A comparison of two theoretical models. *Journal of Management Science*, 35(8), 982–1003.

Dobak, M. and Antal, Z. (2010). *Management and Organisation: Creating and Operating Organisations.* Budapest: Aula Kiada.

Donaldson, L. (2001). *The Contingency Theory of Organisations.* Thousand Oaks, CA: Sage.
Du, S., Bhattacharya, C., and Sen, S. (2010). Maximizing business returns to corporate social responsibility (CSR): The role of CSR communication. *International Journal of Management Reviews*, 12(1), 8–19.
Dubnick, M. (2005). Accountability and the promise of performance: In search of the mechanisms. *Public Performance and Management Review*, 28(3), 376–417.
EFQM (1999). *The EFQM Excellence Model.* Brussels: European Foundation for Quality Management.
Fatoki, O. (2014). The entrepreneurial orientation of micro enterprises in the retail sector in South Africa. *Journal of Sociology and Social Anthropology*, 5(2), 125–129.
Ghalayini, A.M. and Noble, J.S. (1996). The changing basis of performance measurement. *International Journal of Operations and Production Management*, 11(8), 63–80.
Giovannetti, M., Sharma, A., Cardinali, S., Cedrola, E. , and Rangarajan, D. (2022). Understanding salespeople's resistance to, and acceptance and leadership of customer-driven change. *Industrial Marketing Management*, 107, 433–449.
Göpel, M. (2016). Why the mainstream economic paradigm cannot inform sustainability transformations. In M. Göpel, *The Great Mindshift.* Cham: Springer,
Gunasekaran, A., Patel, C., and McGaughey, R. (2004). A framework for supply chain performance measurement. *International Journal of Production Economics*, 87 (3), 333–347.
Heras-Saizarbitoria, I., Marimon, F., and Casadesus, M. (2012). An empirical study of the relationships within the categories of the EFQM model. *Total Quality Management and Business Excellence*, 23(5–6), 523–540.
Hitt, M.A., Ireland, R.D., and Hoskisson, R.E. (2009). *Strategic Management: Competitiveness and Globalisation*, 8th edn. Mason, OH: Cengage Learning.
Hubka, V. and Eder, W. (2012). *Design Science: Introduction to the Needs, Scope and Organisation of Engineering Design Knowledge.* Cham: Springer Science & Business Media.
Iqbal, N., Anwar, S., and Haider, N. (2015). Effect of leadership style on employee performance. *Arabian Journal of Business and Management Review*, 5(5), 1–6.
Ireland, R. and Hitt, M. (2005). Achieving and maintaining strategic competitiveness in the 21st century: The role of strategic leadership. *Academy of Management Executives*, 19(4), 63–77.
Islami, X., Mustafa, N., and Topuzovska Latkovikj, M. (2020). Linking Porter's generic strategies to firm performance. *Future Business Journal*, 6(3),
Ismajli, N., Zekeri, J., Qosja, E., and Krasniqi, I. (2015). The performance of motivation factors on employee performance in Kosovo municipalities. *Journal of Public Administration and Governance*, 5(1), 23–39.
Kabeyi, M. (2019). Organizational strategic planning, implementation and evaluation with analysis of challenges and benefits for profit and nonprofit organizations. *International Journal of Applied Research and Studies*, 5(6), 27–32.
Kaplan, R.S. and Norton, D.P. (1996). *The Balanced Score Card.* Boston: Harvard Business School Press.
Kennerley, M. and Neely, A. (2000). Performance measurement frameworks: A review. In A. Neely (ed.), *Performance Measurement: Past, Present and Future.* Cranfield: Centre for Business Performance.
Kuratko, D. and Audretsch, D. (2009). Strategic entrepreneurship: Exploring different perspectives of an emerging concept. *Entrepreneurship Theory and Practice*, 33, 1–17.

Kyrgidou, L.P. and Spyropoulou, S. (2013). Drivers and performance outcomes of innovativeness: An empirical study. *British Journal of Management*, 24, 281–298.

Lamberg, J., Lubinaite, S., Ojala, J., and Tikkanen, H. (2021). The curse of agility: The Nokia Corporation and the loss of market dominance in mobile phones, 2003–2013. *Business History*, 63(4), 574–605. https://doi.org/10.1080/00076791.2019.1593964

Lengnick-Hall, R., Willging, C.E., Hurlburt, M., and Aarons, G. (2020). Incorporators, early investors, and learners: A longitudinal study of organizational adaptation during EBP implementation and sustainment. *Implementation Science*, 15(74).

Lindelof, P. and Lofsten, H. (2006). Environmental hostility and firm behavior: An empirical examination of new technology-based firms on science parks. *Journal of Small Business Management*, 44(3), 386–406.

Lipczynski, J., Wilson, J., and Goddard, J. (2005). *Industrial Organization: Competition, Strategy, Policy*. Harlow: Pearson Education.

Marques-Quinteiro, P., Curral, L., Passos, A.M., and Lewis, K. (2013). And now what do we do? The role of transactive memory systems and task coordination in action teams. *Group Dynamics: Theory, Research, and Practice, 17*(3), 194–206.

Moura, L., Pinheiro de Lima, E., Deschamps, F., Van Aken, E., Gouvea da Costa, S., Treinta, F., and Cestari, J. (2019). Designing performance measurement systems in non-profit and public administration organisations. *International Journal of Productivity and Performance Management*, 68(8), 1373–1410.

Muchri, P., Pintelon, L., Gelders, L., and Martin, H. (2011). Development of maintenance function performance measurement framework and indicators. *International Journal of Production Economics*, 131(1), 295–302.

Neely, A.D. (1998). *Performance Measurement: Why, What and How*. London: Economics Books.

O'Boyle, I. and Hassan, D. (2013). Organisational performance management examining the practical utility of the performance prism. *Organisation Development Journal*, 31(3), 51.

Osita-Ejikeme, U. (2021). Information technology capability and organisational renewal of hospitality firms in Rivers State. *Research Journal of Management Practice*, 7674.

Paiva, E.L., Roth, A.V., and Fensterseifer, I.E. (2008). Organisational knowledge and the manufacturing strategy process: Aresource-based view analysis. *Journal of Operations Management*, 26, 122–132.

Parijat, P. and Bagga, S. (2014). Victor Vroom's expectancy theory of motivation: An evaluation. *International Research Journal of Business and Management*, 7(9), 1–8.

Park, R.E., Goethert, W.B., and Florac, W.A. (1996). *Goal-Driven Software Improvement: A Guidebook*. Pittsbury: Software Engineering Institute.

Rogers, E.M. (1995). *Diffusion of Innovation*. 4th edn. New York: The Free Press.

Samsami, F., Hosseini, S., Kordnaeij, A. and Azar, A. (2015). Managing environment uncertainty: From conceptual review to strategic management point of view. *International Journal of Business and Management*, 10(7), 215–229.

Sanakulov, N. and Karjaluoto, H. (2015). Consumer adoption of mobile technologies: A literature review. *International Journal of Mobile Communications*, 13(3), 244–275.

Schaltegger, S. and Roger, B. (2017). *Contemporary Environmental Accounting: Issues, Concepts and Practice*. New York: Routledge.

Scott, W.R. (2003). *Organisations: Rational, Natural and Open Systems*. 5th edn. Englewood Cliffs, NJ: Prentice-Hall.

Sigalas, C. and Economou, V. (2013). Revisiting the concept of competitive advantage: Problems and fallacies arising from its conceptualization. *Journal of Strategy and Management*, 6(1), 61–80.

Talbot, C. (2010). *Theories of Performance: Organisational and Service Improvement in the Public Domain*. Oxford: Oxford University Press.

Tapera, J. (2014). The importance of strategic management to business organisations. *Research Journal of Social Science and Management*, 3(11), 122–131.

Taticchi, P. and Balachandran, K. (2008). Forward performance measurement and management integrated frameworks. *International Journal of Accounting and Information Management*, 16(2), 140–154.

Taylor, S. and Todd, P.A. (1995). Understanding Information Technology usage: A test of competing models, *Information Systems Research*, 6(2), 144–176.

Teece, D. (2018). Business models and dynamic capabilities. *Long Range Planning*, 51 (1), 40–49. https://doi.org/10.1016/j.lrp.2017.06.007

Tetrault Sirsly, C. and Lamertz, K. (2008). When does a corporate social responsibility initiative provide a first-mover advantage? *Business & Society*, 47(3), 343–369.

Tuan, L. (2010). Organisational culture, leadership and performance measurement integratedness. *International Journal of Management and Enterprise Development*, 9 (3), 251–275.

Turner, M., Kitchenham, B., Brereton, P., Charters, S., and Budgen, D. (2010). Does the technology acceptance model predict actual use? A systematic literature review. *Information and Software Technology*, 52(5), 463–479.

Turpen, P., Hockberger, P., Meyn, S., Nicklin, C., Tabarini, D., and Auger, J. (2016). Metrics for success: Strategies for enabling core facility performance and assessing outcomes.

Venkatesh, V., Morris, M.G., Davis, G.B., and Davis, F.D. (2003). User acceptance of information technology: Toward a unified view. *MIS Quarterly*, 27(3), 425–478.

Venkatesh, V., Thong, J., and Xu, X. (2012). Consumer acceptance and use of information technology: Extending the unified theory of acceptance and use of technology. *MIS Quarterly*, 36(1), 157–178.

Vinayan, G., Jayashree, S., and Marthandan, G. (2012). Critical success factors of sustainable competitive advantage: A study in Malaysian manufacturing industries. *International Journal of Business and Management*, 7(22), 29.

Wales, W., Covin, J., and Monsen, E. (2020). Entrepreneurial orientation: The necessity of a multilevel conceptualization. *Strategy Entrepreneurship Journal*, 14(4), 633–660.

Part V

The CPS adoption model for facilities management

7 Conceptualised adoption model of cyber-physical systems for facilities management

Introduction

Different disciplines have developed various theories and models depicting the adoption or acceptance of innovative technologies. Venkatesh et al. (2003) stated that the determinants of the adoption of technological innovations are largely influenced by the inherent features or characteristics of the field or discipline under review. With the development of various models and theories for technology adoption or acceptance, such as the Theory of Reasoned Action, the Theory of Planned Behaviour, the Innovation Diffusion Theory, Social Cognitive Theory, the Task Technology Fit Model, the two Technology Acceptance Models, etc. several studies have applied these different theories or models depending on the area or discipline being considered. However, Venkatesh *et al.* (ibid.) observed some limitations in these theories/models after their review and testing, such as:

- the technologies under review were quite simple and also were attributed with individual orientation as against technologies that are more sophisticated and complex;
- the rationale for measurement was cross-sectional;
- most of the studies were conducted in a voluntary set-up, thus enabling a difficult basis for making generalisations.

Against the backdrop of these identified drawbacks, Venkatesh *et al.* incorporated four constructs to determine the model for technology adoption: (1) performance expectancies; (2) social influence; (3) enabling measures; and (4) effort expectancies. These four constructs are used by the current study to determine the adoption of cyber-physical systems (CPS) for facilities management (FM).

As a result of the review of the extant literature on technology adoption, this study considers performance expectancies with 18 variables, social influence with 12 variables, enabling measures with 16 variables, and effort expectancies with 13 variables. As noted earlier, these constructs were used by past studies on technology adoption. Nevertheless, two new

DOI: 10.1201/9781003376262-12

constructs were included in the current study. These are business environment and performance measurement. These two constructs were identified as the gaps in technological adoption concerning facilities management practice. They were deemed important due to their particular roles in the delivery of the mandates of facilities management. These two constructs were attributed with the following: business environment with 12 variables and performance measurement with 16 variables. The aggregate summary of the various constructs (theoretical basis and the included gaps) and their variables is presented in Table 7.1.

Performance expectancy (PEX)

Davis (Davis et al., 1989) outlines performance expectancy as "the degree to which an individual believes that using the system will help him or her attain gains in job performance". Also, several theories and models on technology adoption included performance expectancy as a construct, with some having different nomenclatures. Davis et al. (1992) derived a model called the Motivation Model (MM), having a construct "extrinsic motivation". The study defined this construct as "the perception that users will want to perform an activity because it is perceived to be instrumental in achieving valued outcomes that are distinct from the activity itself, such as improved job performance, pay, or promotions". Also, Thompson et al. (1991) proposed a model entitled the Model of Personal Computing Utilisation (MPCU) and outlined the construct called "job fit". The model defined this construct as how a system's ability can enhance the job delivery of an individual. According to the Innovation Diffusion Theory (IDT), the construct called "relative advantage" indicates the extent to which an innovation is seen as a comparative advantage over a replacement (Al-Rahmi et al., 2019). Anderson et al. (2007) in Social Cognitive Theory outlined the construct of "outcome expectations" in two facets: namely performance expectation and personal expectation.

Social influence (SIF)

In the Unified Theory of Acceptance and Use of Technology (UTAUT) by Venkatesh et al. (2003), social influence is defined as "the degree to which an individual perceives how important others believe he or she should use the new system". Likewise, social influence is also a feature in the constructs of other theories/models on technology adoption. The MPCU model by Thompson et al. (1991) defined social influence as "the individual's internalisation of the reference group's subjective culture and specific interpersonal agreements that the individual has made with others, in specific intentions". While, in Innovation Diffusion Theory, the construct called "image" is outlined as the extent to which an innovation's use is perceived to improve the status or image of an individual in a social system (Moore and Benbasat, 1991). Venkatesh and Davis (2000) in TAM2

Table 7.1 The model constructs and variables

Latent construct	*Measurement variables*
Performance Expectancies (PEX)	Useful for a job task
	Faster accomplishment of tasks
	Increase in productivity
	Reduction of workload
	Performance improvement
	Quality delivery
	Result demonstrability
	Job relevance
	Improvement on the previous system
	Timely delivery of task
	Security of data
	Enhancement of job effectiveness
	Estimation of reliability measures
	Analysis of potential failures
	Asset monitoring and control
	Operations cost optimisation
	Man-hours work reduction
	Effective compliance management
	Health and Safety
Social Influence (SIF)	Organisation's support for the use of the system
	Stakeholders' demand for use
	Utilisation by competing organisations
	Customers' demand
	Elevated prestige
	Increased profile for the organisation
	Competitive pressure
	Network externalities of the organisation
	Influence on organisation's reputation
	Internal network of the organisation
	External network of the organisation
	Compatibility with organisation's cultural values
Enabling Measures (ENM)	Compatibility with work procedures
	Compatibility with the previous system
	Speed and reliability of the system
	Comparison with the previous system

Latent construct	*Measurement variables*
	Training and support
	System's learning flexibility
	Learning capability of personnel
	The convenience of the system's use
	Technology infrastructure
	Resource for procurement of a system
	User satisfaction
	System security
	System stability
	Top management willingness
	Financial resource
	Intellectual resource
	Maintenance costing
Effort Expectancies (EFE)	Self-efficacy
	External support for the use of a system
	System anxiety
	Usability of the system
	Information anxiety
	Skill adaptation
	Ease of use
	Ease of job delivery
	Interaction flexibility
	Users' willingness to use the system
	The complexity of the system
	Compatibility with the organisation's other systems
	System downtimes
Business Environment (BEN)	Organisation's resource commitment
	Work procedure support
	Organisational culture
	Resource availability
	Competitors' innovative strategies
	Business cultural context
	Strategic management capacity
	Service improvement
	Market share

Latent construct	*Measurement variables*
	Competitive hostility
	Environment dynamism
	Business performance
	Business diversification
Performance Measurement (PME)	The built-in capability of facility adaptation
	Identification of problems in facilities
	Improved customer satisfaction
	Customers' involvement in improving the evaluation process
	Informed decision-making
	Significance of cost savings
	Significance of time savings
	Significance of quality assurance
	Attainment of customers' specific needs
	Timely communication of policy changes
	Stakeholders' perception of facilities' performance
	Improvement in facilities' standards
	Economic utilisation of the facility
	Evaluation of existing trends
	Organisation's internal process improvement
	Attainment of future needs of the organisation
Cyber-Physical Systems Adoption for Facilities Management (CFM)	Process optimisation
	Effective task delivery
	Cost efficiency
	Quality assurance
	Time efficiency
	Improved service performance
	Enhanced business performance
	Improved maintenance management
	Space optimisation
	Better life-cycle management
	Effective building adaptation
	Improved workplace coordination
	Attaining business objectives
	Effective capital asset management

(see Chapter 5) described the construct "image" as an individual's viewpoint on people's thoughts that are important to them concerning a particular behaviour.

Effort expectancy (EFE)

The definition of effort expectancy in the Technology Acceptance Model by Davis (Davis et al., 1989) is given as the extent of ease associated with using a system. In other theories/models on the adoption of technology, this is given in different terminologies and defined in various forms. In the Innovation Diffusion Theory (IDT), the construct "ease of use" is defined as the perceived difficulty connected with using an innovation (Moore and Benbasat, 1991). Also, in the MPCU, the construct "complexity" is defined as "the degree to which a system is perceived as relatively difficult to understand and use" (Thompson et al., 1991).

Enabling measures (ENM)

The definition of facilitating conditions in the UTAUT is given as the extent to which an individual's belief in technical or organisational infrastructure exists can serve as an enabler in the functioning of a system (Venkatesh et al., 2003). Furthermore, other theories/models on technology adoption use other terminologies and definitions for facilitating conditions. In the MPCU, the construct "facilitating conditions" is defined as "objective factors in the environment that observers agree to make an act easy to do, including the provision of computer support" (Thompson et al., 1991). The construct "compatibility" in the IDT is outlined as the extent of consistency with laid-down needs, values, and experiences with innovation by likely adopters (Moore and Benbasat, 1991). "Perceived behaviour control" as a construct in the Theory of Planned Behaviour (TPB) is defined as external and internal restraints on perceived reflection on behaviour, which entails self-effectiveness, facilitating conditions for technology adoption and resource-facilitating conditions (Ajzen, 1991).

Model specification and justification

There is a widely accepted view that research on adoption/acceptance and the subsequent use of technology is regarded as one area in the information systems literature that is most mature (Rahi et al., 2019). Yet, there is still the inherent challenge of selecting an ideal or appropriate construct or model from the wide range of models when tasked with the decision to introduce innovative technologies in organisations (Venkatesh et al., 2003). This is associated with the fact that different innovative technologies apply to different disciplines. Hence, validating one model for a particular discipline might not necessarily fit into any other discipline. Brown et al. (2010) noted that there have been different applications of various theoretical models in diverse disciplines to provide an understanding of and make a prediction on the validation of the determinants of technology adoption and usage. Hence, quite a large number of models and theories pose difficulties for researchers when selecting the ideal model for their study.

Consequently, upon reviewing various theories/models on the adoption of technology and the subsequent examination of their strengths and limitations, this research study is premised on the constructs of the Unified Theory of Acceptance and Use of Technology (UTAUT) after being contextualised and modified. The strength and the flexibility of the theory project it as a viable tool in acquiring the views and perceptions of individuals and organisations responsible for facilities management (FM). As a result, the study's conceptual framework examines the underlying strengths of the drivers directing individuals and organisations concerning the adoption of cyber-physical systems for FM. Table 7.2 indicates the constructs that make up the UTAUT and matches their source theories on technology acceptance.

There is a direct correlation between the objectives of the research and the constructs making up the UTAUT. In contextualising the constructs of the theory, four constructs for the adoption of cyber-physical systems for facilities management were considered. These are categorised as: performance expectancy (PEX), social influence (SIF), effort expectancy (EFE), and enabling measures (ENM). Based on the fundamental constructs outlined for technology adoption in previous theories and models, the conceptualised model for this study aims to examine the: relationship between performance expectancy, social influence, effort expectancy, and enabling measures. These constructs are the results of the theoretical basis of this study, hence they form the exogenous variables, likewise their significance in the adoption of cyber-physical systems for FM forms the endogenous variable.

Table 7.2 UTAUT constructs and corresponding technology adoption theory

UTAUT constructs	*Matching construct*	*Matching technology adoption theory*
Performance Expectancy	Perceived Usefulness	Technology Acceptance Model (Davis et al., 1989)
Social Influences	Social Factors	Model of Personal Computing Utilisation (Thompson et al., 1991)
Effort Expectancies	Perceived Ease of Use	Technology Acceptance Model (Davis et al., 1989)
Facilitating Conditions	Facilitating Conditions	Model of Personal Computing Utilisation (Thompson et al., 1991)
Facilitating Conditions	Compatibility	Innovation Diffusion Theory (Moore and Benbasat, 1991)

Source: Adapted from Batarseh (2018).

Structural component of the model

The conceptualised model for the adoption of CPS for FM is based on six exogenous constructs (variables): performance expectancies (PEX), social influence (SIF), enabling measures (ENM), effort expectancies (EFE), business environment (BEN), and performance measurement (PME). The hypothesised model is depicted in Figure 7.1. As indicated in Figure 7.1, the conceptualised model is a multidimensional framework that is made up of the above-mentioned exogenous constructs (variables). The theoretical base is the Unified Theory of Acceptance and Use of Technology (UTAUT) as postulated by Venkatesh et al. (2003).

The UTAUT, as stated, outlines four of the exogenous constructs (variables), which are performance expectancies, social influence, enabling measures, and effort expectancies. However, due to the unique nature of the area of concern (facility management), it was imperative to include particular constructs that best project the intended technology to be adopted (cyber-physical systems); business environment and performance measurement. Hence, the aggregate summary of the entire exogenous variables is carefully crafted to suit its intended purpose, which is the adoption of cyber-physical systems for facilities management.

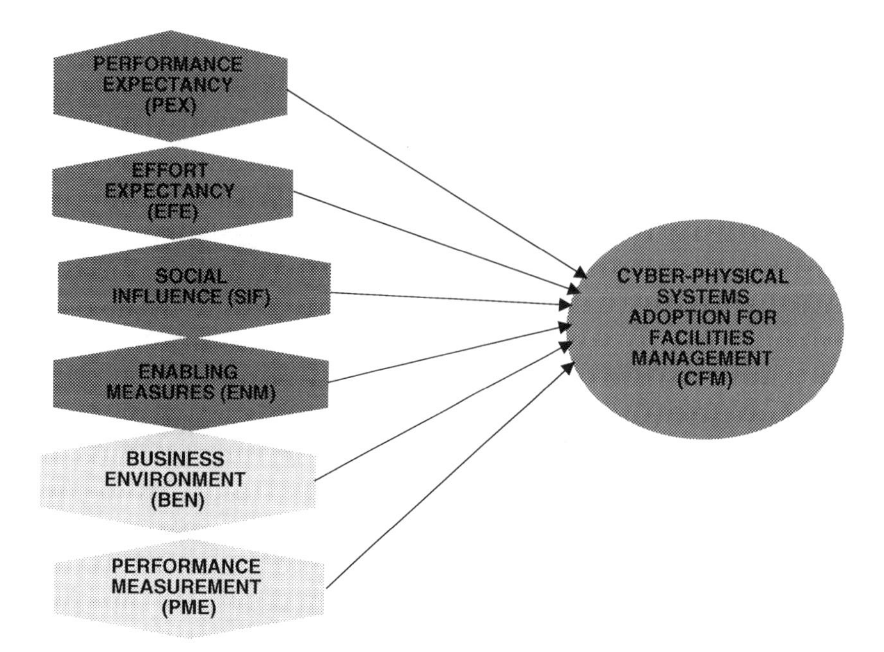

Figure 7.1 Conceptualised cyber-physical systems adoption model for facilities management

Summary

This chapter focused on the conceptualised model for adopting cyber-physical systems for facilities management, which is a multidimensional framework made up of six constructs: performance expectancies, social influence, enabling measures, effort expectancies, business environment, and performance measurement. These variables are products of a productive literature review carried out for the research. Furthermore, a detailed explanation of the theoretical basis for the formulation of the conceptualised model for adopting cyber-physical systems for facilities management was provided.

References

Ajzen, I. (1991). The theory of planned behavior. *Organisational Behavior and Human Decision Processes*, 50(2), 179–211.

Al-Rahmi, W., Yahaya, N., Alamri, M., Aldraiweesh, A., Ali, N., Alturki, U., and Aljeraiwi, A. (2019). Integrating technology acceptance model with innovation diffusion theory: An empirical investigation on students' intention to use e-learning systems. *IEEE Access*, 7, 26797–26809.

Anderson, E., Winett, R., and Wojcik, J. (2007). Self-regulation, self-efficacy, outcome expectations, and social support: Social cognitive theory and nutrition behaviour. *Annals of Behavioural Medicine*, 34(3), 304–312.

Batarseh, S. (2018). Extrinsic and intrinsic drivers of BIM adoption in construction organisations. Unpublished PhD thesis. University of New South Wales, Australia.

Brown, S., Dennis, A., and Venkatesh, V. (2010). Predicting collaboration technology use: Integrating technology adoption and collaboration research. *Journal of Management Information Systems*, 27(2), 9–54.

Davis, F.D., Bagozzi, R.P., and Warshaw, P.R. (1989). User acceptance of computer technology: A comparison of two theoretical models. *Journal of Management Science*, 3(8), 982–1003.

Davis, F.D., Bagozzi, R.P., and Warshaw, P.R. (1992). Extrinsic and intrinsic motivation to use computers in the workplace. *Journal of Applied Social Psychology*, *22*(14), 1111–1132.

Moore, G.C. and Benbasat, I. (1991). Development of an instrument to measure the perceptions of adopting an information technology innovation. *Information Systems Research*, 2(3), 173–191.

Rahi, S., Mansour, M., Alghizzawi, M., and Alnaser, F. (2019). Integration of UTAUT model in internet banking adoption context: The mediating role of performance expectancy. *Journal of Research in Interactive Marketing*, 13(3), 411–435.

Thompson, R.L., Higgins, C.A., and Howell, J.M. (1991). Personal computing: Toward a conceptual model of utilisation. *MIS Quarterly*, 15(1), 125–142.

Venkatesh, V. and Davis, F. (2000). A theoretical extension of the Technology Acceptance Model: Four longitudinal field studies. *Management Science*, 46(2), 186–204.

Venkatesh, V., Morris, M.G., Davis, G.B., and Davis, F.D. (2003). User acceptance of information technology: Toward a unified view. *MIS Quarterly*, 27(3), 425–478.

8 Assessing the conceptualised cyber-physical systems adoption model for facilities management

A Delphi study

Introduction

The book adopted the use of a Delphi study to elicit experts' opinions on the significance and extent of the features of the adoption of cyber-physical systems (CPS) for facilities management (FM), thus, ascertaining the main features that would aid the espousal of CPS for FM. Furthermore, by using experts' opinions, the different types of facilities that would best suit the incorporation of CPS were determined and the extent to which the system would suit such facilities. Also, to predict the future of CPS for FM, experts' opinion was solicited to determine the extent to which the different FM functions would drive the espousal of CPS. A total of two rounds of the Delphi study were conducted to reach a consensus among the experts' views on the questions they answered in the Delphi questionnaire. The chapter includes the background to the Delphi study, the composition of the expert panel, and the analysis of the Delphi results.

The Delphi technique

The Delphi technique is an assessment process that aims to produce a comprehensive assessment and discussion of a specified topic. Buckley (1995) noted that its development originates from the 1950s, when it was used as a problem-solving and forecasting tool at the Rand Corporation. It is named after an old Greek temple housing the oracle at Delphi. According to the ancient Greek myths, the oracle at Delphi was consulted to help forecast the eventual outcomes of significant acts, such as embarking on a war. The Delphi technique was established in the Rand Corporation by Norman Dalkey and Olaf Helmer; and it was used to assess the direction of weapon systems, for war prevention, progress in space research, automation, population control, and scientific breakthroughs (Jones, 1980). Other related studies followed the pattern of the RAND study and subsequently it was deployed by the U.S. Department of Defense, and other government agencies, healthcare agencies, businesses, and factories with the aim of using it as a planning tool and predicting future trends. According to Turoff and Hiltz (1996), researchers at the RAND Corporation

DOI: 10.1201/9781003376262-13

often jokingly referred to the research by Helmer and Dalkey as Delphi research. This was because of how the research was carried out, as the nuclear science experts used anonymity to gain information on future trends in nuclear science and predict the outcomes of the queries.

Falzarano and Zipp (2013) stated that the Delphi technique is a qualitative methodology to produce a consensus among a group of experts on a topic under review. It is a systematic communication method for attaining consensus among a group of experts on an issue of concern (Chan et al., 2001). It refers to a process of iteration in which consensus is attained on a given subject through responses and the opinions of experts usually given in rounds (Hallowell and Gambatese, 2010). Ideally, the selected experts for a Delphi study are experts in the field in the issue being addressed. The technique provides qualitative and quantitative outputs and is attributed with normative, predictive, and explorative elements (Cuhls, 2003). Usually, the process is carried out in rounds (usually two or more), whereby the outcome of the individual rounds is given as feedback to the experts. Hence, the opinion given by the individual experts is based on the outcome of the opinions of other experts. Thus, the Delphi method is a coordinated work system involving group communication among a panel of experts who are geographically dispersed (McMillan et al., 2016). It requires knowledgeable experts whose contribution is collated individually by a singular base (the researcher). At every given round, the researcher analyses the responses from the experts, trying to ascertain central and extreme tendencies, coupled with their validation (Hallowell and Gambatese, 2010). Subsequently, the outcome of the given round is fed back to the experts, whose opinions would then again be collated. The process is continued until a consensus is reached on the issues under review.

Although the Delphi technique is mostly regarded as a qualitative method (Hasson et al., 2000), the technique has received a surge of interest due to the quantitative approach through the combination of quantitative methods (Hallowell and Gambatese, 2010). It is worthy of note that the method is aimed at removing bias that is prevalent when a group of experts physically meet. In the technique, the experts are not furnished with the privilege of getting to know each other. Thus, there is no possibility of an expert "losing face" due to the anonymous nature of the method. It is assumed that the technique brings out the best of group interaction with the use of questionnaires as the basis of interaction (Steurer, 2011). Also, the Delphi technique comes in handy when a long-range forecast is needed since the opinions of the experts are the source of the information that is needed.

It is imperative to state that the Delphi technique has also been criticised over time and supported. One of the leading criticisms of the method is that it is unscientific, with particular emphasis on its lack of psychometric validity (Sackman, 1974) and its level of accuracy (Avella, 2016). Furthermore, Ewing (1979) emphasises that the Delphi technique serves as a last resort when faced with highly complex problems that do not have the right models. At the same time, Fletcher and Marchildon (2014) noted that reliance on the intuitive judgement of the method is a mandatory requirement rather than a temporary expedient. To summarize, as opined by Steurer (2011), the drawbacks of the Delphi technique are: (1) the low level of reliability of the experts' judgements, hence forecasts depend on the selection of judges; (2) resulting sensitivity to the ambiguous nature of the formulated questionnaire deployed for the collection of data for each round; (3) difficulties associated with the level of expertise assessment fused into the forecast. In a distinct review, Haynes and Shelton (2018) compared analysing the Delphi technique and the application of the technique. While affirming that poorly executed Delphi research has been conducted, it was also noted that it would be a fundamental mistake to equate the Delphi method itself with the application of the Delphi method.

Conversely, several studies have supported the use of the Delphi technique. Chang and Ding (2022) noted that there is a firm belief that the Delphi method has great value, not necessarily in the pursuit of public knowledge, but instead in the pursuit of wisdom; not in the pursuit of individual data but in the pursuit of thoughtful judgement. Kauko and Palmroos (2014) showed that the strength of the application of the Delphi technique lies in forecasting financial markets. Furthermore, Helmer (1977) contends that one particular application of the Delphi technique, i.e., the tendency to forecast, is performed in a domain of what might be termed "soft law" and "soft data". Also, Helmer (ibid.) considers that the Delphi method should not be criticised as it is based on opinion and does not follow random sampling rules in the selection of experts. These highlighted criticisms are based on what can be viewed as a gross misunderstanding of the inherent elements of the Delphi technique; hence it is important to point out that the Delphi method should not be confused with an "opinion poll". It is also imperative to state that the ultimate aim of employing the Delphi technique is not to arrive at a majority opinion but rather to attain a consensus or agreement.

Notwithstanding the drawbacks of the Delphi method, as mentioned earlier, Brill et al. (2006) depict the Delphi method as a formidable research method for reaching consensus among experts on a topic of concern, based on the required subjective information and where the selected experts are geographically separated. Moreover, Brill *et al.* (ibid.) noted that the method has been employed in the literature and validated as a reliable approach to empirical studies for attaining consensus in a wide spectrum of fields. These include health care (Whitman, 1990), commerce (Addison, 2003), construction (Aghimien et al., 2020), journalism (Chang and Ding, 2022), education (Yousuf, 2007), and social sciences (Landeta, 2006).

Delphi-specific objectives

The review of the extant literature revealed the different dimensions and attributes related to technology adoption in different fields of human endeavour. Also, several theories and models have been postulated to inform the adequacy and preparedness of adopting various technological innovations over the years. However, after exhaustively reviewing the literature on the theories above and the models for technological innovation adoption, there do not appear to be any existential studies related to how the adoption of CPS for FM can be achieved. It is important to note that CPS serve as one of the conveyors of the Fourth Industrial Revolution. Hence, it is expected that such sophisticated technological innovation is very unlikely in most countries.

Considering the importance of facilities management to the built environment, it is pertinent to develop a gateway that will serve as a base for the adoption of CPS for FM. The use of the Delphi technique helped test the attributes identified from the literature that were considered significant in the drive for the adoption of CPS for FM. From the opinions of the Delphi experts in the study, the main attributes and the sub-attributes that were considered to encourage the adoption were ascertained, based on their significance. However, the ultimate aim was to assess and validate the conceptual model which was hypothesised in Chapter 7 after an exhaustive review of theories and models regarding technology adoption.

With the aim of the Delphi study established as developing a conceptual model for the adoption of CPS for FM, the following were set as the specific objectives for the Delphi study:

- **DSO1** To identify the main features for adopting cyber-physical systems for facilities management.
- **DSO2** To identify the sub-features for adopting cyber-physical systems for facilities management.
- **DSO3** To determine the extent to which cyber-physical systems can be adopted for the different types of facilities.
- **DSO4** To predict the extent to which the various functions of facilities management can serve as drivers for the adoption of cyber-physical systems.

Selection of panel experts

To a great extent, the ultimate success of a Delphi study is hugely dependent on the objective and careful selection of the experts making up the panel (Chan et al., 2001). The experts constituting the panel must have an adequate interest in the topic and have a considerable stake in the subject of concern. This will ensure a high committal rate on the part of the experts and ensure that the expertise of the panellists offers immense advantage to the process. Hallowell (2008) stated that the experts engaged in the Delphi process are notably professionals or researchers who have vested experience and knowledge and are reflected through a range of specific requirements, including work experience, professional qualification, job appointment, and relevant publications.

However, various schools of thought have differed on what an "expert" is. Hasson et al. (2000) noted that controversy arises when a professional is projected as an "expert". This book adopts the definition of an expert as outlined by McKenna (1994) to get a clearer picture of the subject. The study connotes an "expert" panel as being a panel of informed individuals. This is also stated by Goodman (1987, p. 730): "the Delphi technique tends not to advocate a random sample of panellist … instead, the use of experts or at least of informed advocates is recommended". Furthermore, Helmer (1977, p. 18) contends that

> since a Delphi inquiry is not an opinion poll, relying on drawing a random sample from the population of experts is not the best approach, rather, once a set of experts has been selected (regardless of how – but following predetermined qualifying criteria) – it provides a communicative device for them that uses the conductor of the exercise as a filter to preserve the anonymity of the responses.

Hence, the most significant risk in the selection of panellists is to take the route of "least resistance" by selecting a group of individuals who are like-minded or cosy friends, which eventually decreases the strength of the Delphi process.

Hemmat et al. (2021) advise the inclusion of key experts in the area of concern in order to obtain the latest thinking and give directions for policy-making. They also recommend the integration of experts' opinions with existing ones for the formulation of a research base for policy-making. However, the potential for bias is very likely when selecting experts for the panel. Hasson et al. (2000) emphasise the need for neutrality and impartiality when selecting the experts to make up the Delphi panel. However, this might not be entirely eliminated as experts might exhibit tendencies of misconception. A range of explicit selection criteria should be outlined as a basis for evaluating results and establishing the study's potential importance in relation to other set-ups and the population, since panel experts serve as the cornerstone of the Delphi process (Iqbal and Pipon-Young, 2009). Niederberger and Spranger (2020) opined that several criteria can be considered in the selection process for panel experts. Notably, two recommendations were canvassed. First, is the need for the experts to exhibit a resounding knowledge base of the issue under review. Second, the experts should serve as representatives of the profession to ensure seamless acceptance of their suggestions by the population. In a similar projection, Adler and Ziglio (1996) advocate four necessary requirements to be met by panel experts in a Delphi study: (1) experience in and knowledge of the subject of concern; (2) willingness and capacity for participation; (3) enough time for the entire process; and (4) coherent and effective communication skills.

For this book, the selection of experts for the panel of the Delphi technique was carried out in conformity with the attainment of at least five of these criteria:

1 *Knowledge*: The individuals must have sufficient knowledge of the field of information and communication technology, knowledge of the field of facilities management, and knowledge of construction management.

2. *Academic qualification*: The individuals must have acquired a degree (National Diploma, Bachelors, Masters, PhD) in facilities management/ construction management, most notably specialising in information and communications technology.
3. *Experience*: The individuals should have a background of having worked in the construction industry or being involved in the management of facilities. This can be with consultancy organisations, contracting organisations, or government establishments.
4. *Employment*: The individuals must be currently serving or have served as professionals in any institution, agency, department, business, organisation, or company.
5. *Authorship*: The individuals should have authored or co-authored research publications in peer-reviewed journals in construction management or facilities management; made presentations at conferences, symposiums, professional meetings, or workshops.
6. *Recognition*: The individuals must have served as a peer-reviewer for at least one manuscript from the editor of a revered journal with an emphasis on facilities management.
7. *Teaching*: The individuals must have carried out teaching lessons or classes, notably focusing on facilities management, and made presentations at workshops that aim to proffer professional or advisory, or technical expertise in the area of facilities management.
8. *Research*: The individuals must have been engaged in research functions or duties or must have submitted a minimum of one research proposal or must have gained a research grant or funding from a funding agency (government/private) in the area of facilities management.
9. *Professional membership*: The individuals must be a member of a professional body (most notably in the area of construction or facilities management or information technology).
10. *Willingness*: The individuals must exhibit the unreserved willingness to participate in the whole Delphi process.

The expert panel was recruited through emails that contained a general overview of what the study was all about. Also, their consent for inclusion in the Delphi study was solicited, and those whose participation was confirmed were sent a comprehensive description of the Delphi process. Furthermore, a request was made for the curriculum vitae (CV) of the confirmed experts so as to ascertain their area of expertise and also to assess their conformity with the stipulated qualifying criteria. After verifying their CVs, it was affirmed that all the experts selected for the study met the qualifying criteria for the study. A total of 29 invitations were sent out, while 11 responses were received. This was followed by sending out the initial round of the questionnaire survey. The questionnaire had both closed and open-ended questions. The 11 experts making up the panel were retained for both the first and second rounds of the Delphi process. This was deemed adequate since there is no significant correlation between a Delphi panel size and the efficiency and accuracy of the Delphi method (Balasubramanian and Agarwal, 2012). Based on the outlined recommendations made by the studies above, coupled with the notion that the efficacy of the Delphi technique is not wholly dependent on statistical power, it was felt that, instead, its effectiveness is tailored towards the waves of attaining a consensus among a group of experts. Hence, a Delphi panel size of 11 members was deemed suitable.

Establishing the panel size

According to Vernon (2009), there is no agreed unanimity concerning the Delphi study panel size, nor a clear definition or recommendation of "small" or "large" samples. There is generally a lack of agreement on the sample size for expert panels and the generally accepted criteria upon which the judgement of the choice of sample size is made. A huge pool of different studies has adopted virtually any sample size. Reid (1988) observed that various published articles state finding 10–1685 panellists in different fields that employed the Delphi technique. Hence, it can be argued that different scholars have recommended using a wide range of sample sizes. For example, Strasser et al. (2005) adopted the use of six experts in the study undertaken to identify the seriousness and likelihood of drug interaction in ambulatory pharmacies.

Furthermore, Harder et al. (2010) advocated for the use of 12 panellists in the conduct of a Delphi study. Howell and Kemp (2005) employed 13 experts in research that involved a different spectrum of skills exhibited by young children. In the study to propose recommendations for treating gastro-oesophageal diseases, Armstrong et al. (2005) used a panel of 23 experts in the adopted Delphi technique. Moreover, Dunn (1994) proposed the use of 10–30 experts, explaining that as there is an increase in the complexities of policies, there should be a corresponding larger sample size to include a vast range of opinions which should represent for and against the area of concern. Dunn (1994) further states that formal and informal participants should be included, based on their interest in the subject matter. These participants should be affiliated with different groups, holding several positions, and having a variety of extent of influence; this is the supposition upon which this current study was based.

For studies that have used the Delphi technique, the adoption of sample size has been seen to be researcher-driven and situation-specific. In contrast, the employment of convenience samples has been mostly used depending on the availability of the experts and the resource. Due to the lack of a generally agreed standard sample size in the Delphi technique, the verdict on how small a panel size is cannot be overtly stated. If there is homogeneity among the experts, then 10–15 participants ought to be sufficient (Andranovich, 1995). If the groups exhibit diverse interests, there should be an increase in the number of participants (Harder et al., 2010). In determining the sample size of a Delphi study, Skulmoski et al. (2007) underscored the following factors that should be considered:

1 *Homogeneous or heterogeneous sample*: A sample size of 10–15 participants should be deployed when the group is homogeneous to produce good results. However, when a heterogeneous group is involved, it is advocated that a larger sample should be used.
2 *Quality of decision*: As there is an increase in the sample size, there is a corresponding increase in the quality of the decision (or group error reduction). Nevertheless, when it is beyond a given threshold, the management of the Delphi process becomes overwhelmingly complex.
3 *External and internal verification*: The credibility of the confirmation of the outcome of the Delphi process is better attained with a larger group, although the result from the use of a small group can be verified with follow-up research. The use of a small group was applied for this study, and the verification will be conducted with a questionnaire survey.

The Delphi process

Delphi iterations

The Delphi process involves preparing the questionnaire, analysing the response, and the respondents' feedback, which are carried out in rounds until the attainment of the desired consensus. Suris and Akre (2015) noted that sequential rounds of questionnaire distribution and retrieval are deployed to obtain iterative responses to the subject of concern. However, there is no generally agreed number of rounds that should be engaged in during the Delphi technique. Kloser (2014) suggested three rounds; while Kim et al. (2022) suggested that the appropriate number of rounds for a Delphi study should be two or ten rounds, insisting that if the number of rounds increases, higher accuracy is achieved. Although, Powell (2003) emphasised that the higher the number of rounds engaged in, the more time-consuming it will be, with higher susceptibility to fatigue, and a higher rate of attrition. This study aimed at achieving consensus after the second round of the Delphi study by the selected experts of the panel on the attributes for the adoption of CPS for FM. A well-detailed and comprehensive questionnaire focused on the issues pertaining to the adoption of CPS for FM was sent to the Delphi participants. The formulation of the questionnaire was the outcome of a rigorous and detailed review of the extant literature, coupled with the review of theories and models on technology adoption. The questionnaires were emailed to the individual expert members.

As earlier noted, the current study had an iterative process of two rounds. Individual rounds were completed on average in one month, with the experts being allowed sufficient time to make a quality contribution. The first-round questionnaire was solely based on the literature review findings, while that of the second round was based on the responses from the experts from the first round. This means that the second round of the Delphi process was the result of the brainstorming exercises of the involved panellists given in the first round. The first round had closed and open-ended questions, which gave the experts a means of offering new ideas, culminating in the formulation of the second round of questions. In both rounds, the responses were analysed to ascertain the degree of consensus from the elicited experts' opinions. The second round of the questionnaire distribution allowed the experts to engage in a review of the attributes relating to the adoption of CPS for FM, which the researcher did not propose in the first round. In the second round, closed questions were deployed to ascertain the participating experts' agreement on the attributes (which also included the newly introduced attributes). These responses were also further subjected to analyses to ascertain whether consensus has been obtained on the issues under review. After the analyses were carried out on the elicited responses from the experts in the second round, virtually all the attributes and other related issues had attained consensus. Findings from the Delphi study culminated in the validation of the study's conceptual framework, which was further subjected to validation through a questionnaire survey.

Achieving consensus in the Delphi study

One of the motives for carrying out a Delphi study is the determination of consensus using the opinions of the panel of experts on the subject under review. For the current study, establishing a consensus on the identified main attributes and sub-attributes was imperative. In determining the consensus on the questions posed to the experts, the study used the group median and the interquartile deviation (IQD). Hence, a consensus was achieved among the various questions if the deviation among the group median of the responses did not exceed one unit. Also, this was employed for the IQD of the individual questions. For the IQD, it entails arriving at the absolute difference between the 75th and 25th percentiles. A percentile serves as a representation of values of tests that is non-referenced; the first quartile (Q1) is referred to as the 25th percentile, the second percentile (Q2) is referred to as the 50th percentile and also the median, and the third quartile (Q3) is referred to as the 75th percentile. In arriving at the IQD, the difference between the 75th and 25th percentiles is computed. When the IQD value is small, it denotes a great extent of agreement or consensus; while if the IQD value is high, it denotes a great extent of disagreement or dissensensus.

The determination of consensus on a given set of opinions is a difficult process to actualise. No generally accepted rule establishes how consensus should be attained in the conduct of a Delphi study. Since consensus refers to the agreement, it can thus be attained through the cumulative waves of judgements, geared towards a subjective stance of the central tendency, or confirmation of the opinion stability in consonance with consistent views in the various rounds of the process (Kim et al., 2022).

Various studies have deployed several statistical techniques in determining consensus for the Delphi process. Rayens and Hahn (2000) adopted means and standard deviations, whereby a decrease in standard deviation in successive rounds indicated a high tendency for consensus. McKenna (1994) projected frequency distribution in assessing consensus with the set criterion of 51 per cent. Also, some studies have used IQD to establish consensus (Aigbavboa, 2013). Furthermore, Olij et al. (2017) noted that an IQD less than or equal to 1.00 indicates the attainment of consensus, while a difference of more than 1.00 IQD value obtained at a continuous stage portrays the convergence of views, a consensus is achieved (Aigbavboa, 2013). Moreover, Holey et al. (2007) outlined the following criteria for the determination of consensus:

1 Percentage response.
2 Individual questions: the percentage for their level of agreement for the process of different rates of response.
3 Analysis of the questions' median, standard deviation, and the accompanying group rankings.
4 Analysis of the questions' means, standard deviation, and the accompanying group rankings.
5 Analysis of the weighted kappa (k) values for the comparison of round's chance eliminated agreement.

Furthermore, Holey et al. (ibid.) presented the following to indicate when consensus is achieved:

1 Increment in the percentage of agreement.
2 Convergence of importance rankings.
3 Kappa values increase.
4 Reduction in standard deviation values.
5 Smaller range of responses.
6 A reduction in comments proffered in the progression of rounds.

As earlier noted, there are no generally agreed criteria for measuring consensus in a Delphi process. However, a common denominator among all the postulated ideas is that there must be a level of convergence of opinions among the panel of experts. This is bolstered by the alignment of their ideas towards a subjective measure for a central tendency. For the attainment of consensus in the Delphi study for this thesis, the following criteria were set:

1 An average of 60 per cent of the responses are either positive or negative on the provided questions.
2 Not exceeding one unit of the absolute deviation of the provided questions.
3 An IQD value that is less than 1.00. Thus, values of IQD = 1 portray a very high consensus.

Hence, the current study adopted the following scales for the categorisation of consensus:

1 Strong consensus – median 9–10, mean 8–10, IQD ≤1 and ≥ 80 per cent (8–10).
2 Good consensus – median 7–8.99, mean 6–7.99, IQD ≥1.1 ≤2 and ≥ 60 per cent ≤ 79 per cent (6–7.99).
3 Weak consensus – median ≤ 6.99, mean ≤ 5.99, IQD ≥2.1 ≤3 and ≤ 59 per cent (5.99).

Result from the Delphi study

Background information of experts

A total of 11 experts took part in both rounds of the Delphi study. In choosing the panel members, particular importance was placed on the experts' background which had to include experience in construction processes and management that is technologically inclined. This was deemed important because of the nature of the research, which focused on the management of built facilities using cyber-physical systems. Furthermore, to strike a balance between theorists and practitioners, the selection of expert panel members was all-inclusive of both backgrounds.

Table 8.1 shows the background information of the experts with regard to their attainment of the selection criteria as stipulated for this study. Nine of the experts were male, while two were female. Based on the academic qualification of the selected experts, it was revealed that four of the experts possessed a Bachelor's degree as their highest professional qualification, thus making up 36.36 per cent of the total number; while three have a Master's degree, which indicates 27.28 per cent of the total number, and four of the experts have a Doctorate, which makes up 36.36 per cent of the total number (Table 8.1). Furthermore, concerning the area of specialisation of the selected experts, four specialised in FM, four in construction management, two in building construction, and one in information technology. Based on the years of professional experience of the experts, one had 1–5 years of professional experience, while three had 6–10 years and five had 11–15 years; and one expert had 16–20 years and one had over 20 years. Moreover, based on the employment agency of the selected experts, three were affiliated with consultancy establishments, three worked with contracting organisations, and government agencies employed five.

Figure 8.1 presents the publication credentials of the selected experts for the Delphi study. Eight of the experts had published in both peer-reviewed journals and peer-reviewed conference proceedings. Five had published book chapters, while seven had served as reviewers for journals and eight had served as reviewers of conference proceedings. Also, all of the experts had been engaged in formulating technical reports, while three had served as members of the editorial board of reviewed journals.

The Delphi questionnaire served as the instrument for eliciting responses from the expert members. Individual rounds had their own particular questionnaire,

Table 8.1 Summary of the experts' background

Category	*Classification*	*No. of experts*	*Percentage*
Academic qualification	Bachelor's degree	4	36.36
	Master's degree	3	27.28
	PhD	4	36.36
	Total	11	100.00
Area of specialisation	Facilities management	4	36.36
	Construction management	4	36.36
	Building construction	2	18.19
	Information technology	1	9.09
	Total	11	100.00
Years of experience	1–5 years	1	9.09
	6–10 years	3	27.28
	11–15 years	5	45.45
	16–20 years	1	9.09
	Over 20 years	1	9.09
	Total	11	100.00
Employment agency	Consultancy	3	27.27
	Contractor	3	27.27
	Government	5	45.46
	Total	11	100.00

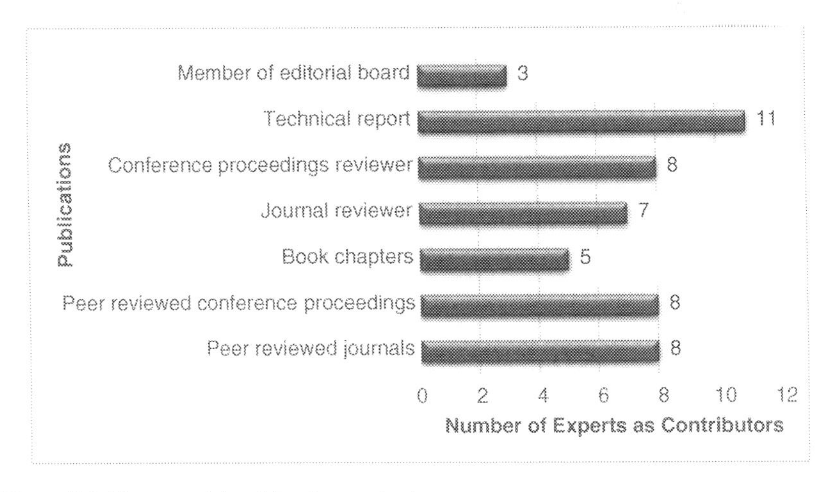

Figure 8.1 The experts' publication output

albeit that of the second round was based on the responses from the initial round. The design of the questionnaire for round one was based on the identification of attributes (main one and sub-attributes) from the comprehensive review of the extant literature. The identified attributes from the reviewed literature were constructively and systematically coupled to formulate the initial round of the Delphi survey. This initial round, therefore, served as a brainstorming process to produce an itemised list of significant features that pertain to the adoption of CPS for FM to actualise the outlined objectives. The first round had open-ended questions, which gave the experts the platform to include features they deemed significant in the category under review. The outcome of the first round formed the basis for the formulation of the second round of the study. The degree of consensus attained by the responses given by the experts with respect to the adoption of cyber-physical systems for FM was ascertained with the appropriate statistical tools, median and the IQD.

The aim of the second round of the Delphi survey was to enable expert members of the panel to review, critique, and comment on the features of the adoption of cyber-physical systems for FM that was initially proposed by the experts in the initial round of the Delphi survey. This round had closed questions that examined the experts' reactions in terms of agreement, disagreement, or explanation regarding the proposed features of the adoption of CPS for FM.

In round two of the Delphi survey, the opinions of experts were generally all in conformity. This portrayed a successful alignment of all questions on the features relating to the adoption of CPS for FM; hence there was no need for a third round of the survey. In the individual rounds, the mean, median, standard deviation, and IQD of the individual questions were calculated, which enabled attainment of consensus or not. In scenarios where the value was two points above or below the median value, the experts were asked to give reasons for such a choice. The experts' opinions converged on the individual questions making up the Delphi survey.

DSO1 To identify the main features for the adoption of cyber-physical systems for FM

After a comprehensive review of the extant literature, the attributes indicating the features that could serve as significant to the adoption of CPS for FM were identified. The outlined features were presented to the experts to determine their significance in the discussion of the espousal of CPS for FM. Presented on an ordinal scale, the experts were asked to rank the individual features on a scale of 1–10, with 1 indicating no significance and 10 indicating high significance.

The study identified six main features for adopting CPS for FM. Five of these features were considered by the experts to be of very high significance (VHS) based on the median score of 9.0 and 10.0 (VHS:9.00–10.00). As shown in Figure 8.2, only one feature had a high significance (HS) with a median score of 8.00 (HS: 7.00–8.99). Also, none was found to have medium

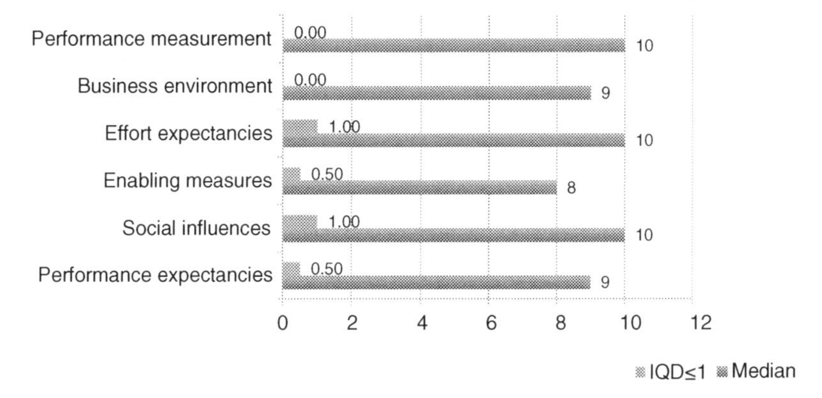

Figure 8.2 Significance of the main features for the adoption of CPS for FM

significance (MS) (MS: 4.00–6.99). Equally, in the opinion of the experts, none was found not have any significance in the determination of the adoption of CPS for FM. The IQD scores derived from the experts' opinions show that all the features attained a strong consensus, with all the features having values whose IQD is within the premise of the study's cut-off of less than or equal to 1.00.

DSO2 To identify the sub-features for adopting cyber-physical systems for facilities management

Through the comprehensive review of the extant literature, the main features peculiar to the espousal of CPS for FM were identified and served as the main constructs in the conceptualised model. Furthermore, these main features all have sub-features that measure the identified underlying constructs (main features). These sub-features were also presented to the expert panel of the Delphi study to outline their significance in the adoption of CPS for FM. This was carried out using the ordinal scale from 1–10, with 1 being very low significance and 10 being very high significance.

A total of 19 variables made up the first sub-feature list (performance expectancies). As shown in Figure 8.3, it is indicated that three had a value of very high significance (VHS: 9.00–10.00), whereas 16 were shown to have a high significance (HS: 7.00–8.99). Conversely, none of the variables was shown not to have any significance on the adoption of CPS for FM. Furthermore, the IQD scores of the variables indicate that consensus was attained for all the items (19), with all having values between 0.00 and 1.00. This shows consistency among the panel of experts on the variables making up the construct (performance expectancies) to be of very viable significance in the discussion on adopting CPS for FM.

Regarding social influences, a total of 14 variables made up the sub-feature. Four of the variables were deemed to be of very high significance (VHS: 9.00–10.00) in

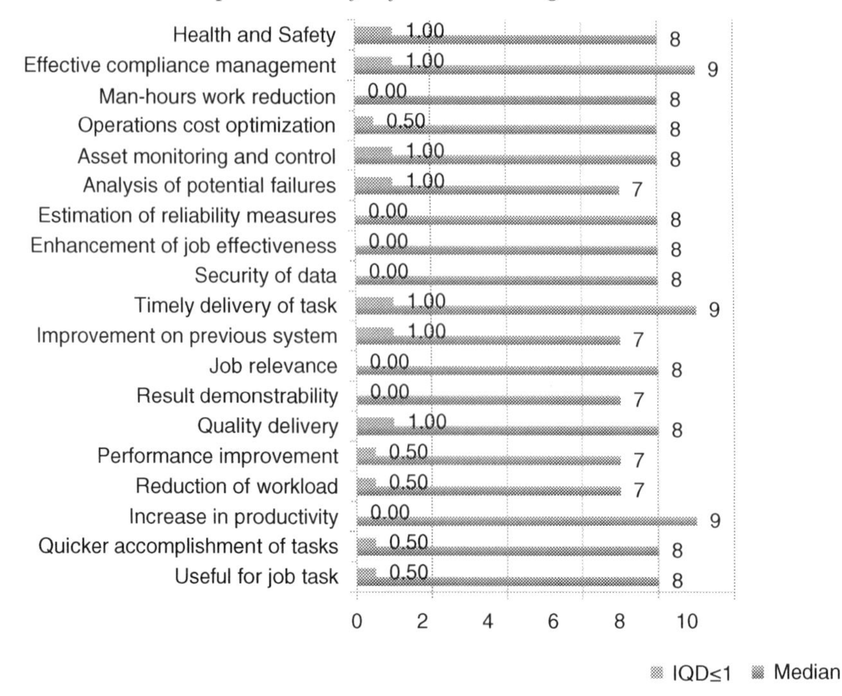

Figure 8.3 Significance of performance expectancies on the adoption of CPS for FM

the determination of the adoption of CPS for FM, while ten were adjudged to be of high significance (HS: 7.00–8.99). Also, none of the variables was outlined not to be of significance as indicated in Table 8.2. Furthermore, using the IQD value of the variables, all variables were deemed to attain consensus except two. These are used by better-performing organisations as a status symbol for the organisation. Also, the use by better-performing organisations had a mean rating of 8.45 and an SD of 1.13, while the status symbol for the organisation had a mean rating of 8.45 and an SD of 1.63.

The third sub-feature is enabling measures and has 18 variables. As shown in Table 8.3, six of these variables were deemed to be of very high significance (VHS: 9.00–10.00) by the panel of experts in the determination of the adoption of CPS for FM; while 12 were adjudged to be of high significance (HS: 7.00–8.99). Conversely, none of the variables was found to be of no significance in determining the adoption of CPS for FM. Furthermore, the IQD values of the variables all achieved consensus with a value of 0.00–1.00 except for one, which is interactiveness with the system with an IQD value of 1.5. Moreover, interactiveness with the system had a mean rating of 8.82 and an SD of 0.98.

The fourth sub-feature is effort expectancies and had 15 variables. As indicated in Table 8.4, four of the variables were deemed to be of very high significance (VHS: 9.00–10.00) by the panel of experts in the determination of

Table 8.2 Social influence

Features	*Median*	*Mean*	*SD*	*IQD*
Organisation's support for use of the system	9	9.09	1.14	0.00
Stakeholders' demand for use	8	8.82	0.98	0.00
Used by competing organisations	8	8.64	0.92	1.00
Customers' demand	9	9.09	0.70	0.50
Elevated prestige	7	8.55	1.37	1.00
Increased profile for the organisation	9	9.09	0.54	0.00
Status symbol for the organisation	7	8.45	1.63	1.50
Competitive pressure	9	8.73	0.90	1.00
Network externalities of the organisation	7	8.45	1.13	0.50
Use by better-performing organisations	7	8.45	1.13	1.50
Influence on organisation's reputation	8	8.91	1.04	0.50
Internal network of the organisation	8	8.73	0.90	0.00
External network of the organisation	8	8.82	1.08	0.00
Compatibility with organisation's cultural values	7	8.55	1.37	1.00

Note: SD = standard deviation; IQD = interquartile deviation.

Table 8.3 Enabling measures

Features	*Median*	*Mean*	*SD*	*IQD*
Compatibility with work procedures	8	9.09	0.54	0.00
Compatibility with previous system	7	8.55	1.13	1.00
Speed and reliability of system	9	9.45	0.52	1.00
Comparison with previous system	8	8.82	1.08	0.00
Training and support	8	9.18	0.60	0.50
System's learning flexibility	7	8.73	0.90	1.00
Learning capability of personnel	8	9.09	0.70	0.50
Convenience of the system's use	8	9.00	1.10	1.00
Interactiveness of the system	8	8.82	0.98	1.50
Technology infrastructure	9	9.00	0.98	0.50
Resource for procurement of system	8	9.18	0.40	0.00
User satisfaction	9	9.27	0.47	0.50
System security	9	9.27	0.65	1.00
System stability	9	9.27	0.65	1.00
Top management's willingness	8	9.09	0.83	0.50
Financial resources	9	9.36	0.67	0.00
Intellectual resources	7	8.64	0.92	0.00
Maintenance costing	8	9.00	1.00	1.00

Note: SD = standard deviation; IQD = interquartile deviation.

Table 8.4 Effort expectancies

Features	*Median*	*Mean*	*SD*	*IQD*
Self-efficacy	7	8.16	1.33	0.00
External support for use of system	9	8.64	0.67	0.50
System anxiety	7	7.73	0.90	0.50
Usability of the system	9	8.82	0.87	0.00
Information anxiety	7	7.64	0.67	1.00
Skill adaptation	8	8.36	0.92	0.50
Ease of use	9	8.82	0.87	1.00
Ease of job delivery	8	8.55	1.21	1.00
Interactiveness flexibility	8	8.45	0.93	0.50
Users' willingness to use the system	7	8.00	1.26	0.00
Complexity of the system	8	8.27	0.79	0.00
Compatibility with organisation's other systems	8	8.45	0.82	0.00
User comfort	8	8.45	1.29	1.50
Ease of learning	9	8.91	0.83	1.50
System downtimes	8	8.64	0.81	1.00

Note: SD = standard deviation; IQD = interquartile deviation.

the adoption of CPS for FM, while 11 are adjudged to be of high significance (HS: 7.00–8.99). Conversely, none of the variables was found to be significant in the determination of the adoption of CPS for FM. Furthermore, IQD values of the variables all achieved consensus except for two of the variables: ease of learning and user comfort. Both variables have an IQD value of 1.5. Moreover, the mean rating of ease of learning is 8.91 and SD is 0.83, while user comfort had a mean rating of 8.45 and SD of 1.29.

The next sub-feature is the business environment which had 16 variables. As shown in Table 8.5, eight of the variables are considered to be of very high significance (VHS: 9.00–10.00) by the panel of experts in the determination of the adoption of CPS for FM, while eight variables are deemed to be of high significance (HS: 7.99–8.00). Conversely, none of the variables were found to be of no significance in determining the adoption of CPS for FM . Also, IQD values of the variables all achieved consensus except for three variables: market orientation, return on investment, and sales growth. Moreover, the mean rating of market orientation is 7.82 and the SD is 1.01; while return on investment had a mean rating of 8.86 and an SD of 0.89; and sales growth had a mean rating of 8.73 and an SD of 1.01.

The sixth sub-feature is performance measurement and it had 16 variables. As indicated in Table 8.6, nine of the variables are considered to be of very high significance (VHS: 9.00–10.00) by the panel of experts in the determination of the adoption of CPS for FM, while seven variables are deemed to be of high significance (HS: 7.99–8.00). Conversely, none of the variables was

Table 8.5 Business environment

Features	*Median*	*Mean*	*SD*	*IQD*
Organisation's resource commitment	9	8.82	0.87	0.00
Work procedure support	8	8.18	0.98	0.00
Market orientation	9	8.73	1.01	1.50
Organisational culture	8	7.82	0.98	0.50
Resource availability	9	8.73	0.79	0.00
Competitors' innovative strategies	8	8.27	1.19	1.00
Business cultural context	8	7.73	1.10	0.00
Strategic management capacity	9	8.82	0.87	1.00
Return on investment	9	8.86	0.89	2.00
Sales growth	9	8.73	1.01	1.50
Service improvement	8	8.55	1.04	1.00
Market share	8	8.55	0.93	1.00
Competitive hostility	8	8.45	0.82	0.50
Environment dynamism	8	8.55	0.93	0.00
Business performance	9	8.73	1.01	1.00
Business diversification	9	9.00	1.00	1.00

Note: SD = standard deviation; IQD = interquartile deviation.

Table 8.6 Performance measurement

Features	*Median*	*Mean*	*SD*	*IQD*
Built-in capability of facility adaptation	8	8.64	1.03	0.00
Identification of problems in facilities	8	8.64	1.12	0.00
Improved customer satisfaction	9	8.91	1.14	0.00
Customers' involvement in improving evaluation process	8	9.09	0.54	0.00
Informed decision-making	8	8.82	0.75	1.00
Significance of cost savings	10	9.55	0.52	1.00
Significance of time savings	9	9.27	0.65	1.00
Significance of quality assurance	9	9.18	0.75	1.00
Attainment of customers' specific needs	9	9.18	0.40	0.00
Timely communication of policy changes	9	8.91	0.83	0.00
Stakeholders' perception of facilities performance	9	9.09	0.83	0.50
Improvement in facilities' standards	9	9.18	0.60	0.50
Economic use of the facility	8	8.73	0.90	1.00
Evaluation of existing trends	8	9.09	0.83	0.50
Organisation's internal process improvement	9	9.18	0.60	0.50
Attainment of future needs of the organisation	8	8.82	0.75	1.00

Note: SD = standard deviation; IQD = interquartile deviation.

found to be significant in the determination of the adoption of CPS for FM. Also, the IQD values of the variables all attained consensus.

DSO3 To determine the extent to which cyber-physical systems can be adopted for the different types of facilities

The different types of facilities all have their intended functions and uses depending on the set-out plans and ideas for which they are initially constructed. This ranges from residential houses, industrial facilities and manufacturing layouts, healthcare infrastructure, which caters to healthcare delivery, short-term or long-term storage facilities to house and store unfinished/or finished goods, to religious buildings and hospitality facilities, such as hotels or spas, etc.

It is imperative to point out that the espousal and use of cyber-physical systems in the different types of facilities will be based on varying dimensions. Since the various types of facilities all have different uses/functions, the adoption of cyber-physical systems might help more in the optimisation of process delivery in some types of facilities as compared to others. Adopting cyber-physical systems might bring comparative advantage in some types of facilities, such as monitoring physical assets, management of the workflow process, overall supervision of the building facilities, and maintenance management. Some of the types of facilities would be better suited to incorporate the lofty system compared to others. Given the aforementioned, it is important to discover the extent to which cyber-physical systems can be used in the different types of facilities for the delivery of their functions.

An identification of 13 types of facilities was made by the study. This list was put to the panel of experts to ascertain the extent to which the incorporation of cyber-physical systems can aid the delivery of their functions. As indicated in Table 8.7, experts believe that industrial and healthcare facilities are best suited for the espousal of cyber-physical systems having (Mean = 9.18; SD = 0.98, IQD = 0.00) and (Mean = 9.18; SD = 0.98; IQD = 0.00) respectively. The third-ranked facility is retail, having (Mean = 8.91; SD = 0.83; IQD = 0.00); while the fourth-ranked are hospitality and educational institutions, having (Mean = 8.82; SD = 0.63; IQD = 1.00) and (Mean = 8.82; SD = 0.87; IQD = 8.82) respectively. Furthermore, it was revealed that all types of facilities attained consensus except two: residential and religious buildings.

DSO4 To predict the extent to which the various functions of facilities management can serve as drivers for adopting cyber-physical systems

Facilities management entails a wide range of operations and activities, ranging from areas, such as the management of space and property, the control of the environment, health and safety, control and monitoring of various units of the organisation, and support services. Other functions include financial management, management of real estate, building maintenance, and asset management. These functions serve as the basis for the importance of

Table 8.7 Types of facilities

Features	*Mean*	*SD*	*IQD*	*Rank*
Industrial	9.18	0.41	0.00	1
Healthcare	9.18	0.98	0.00	2
Retail	8.91	0.83	0.00	3
Hospitality	8.82	0.63	1.00	4
Educational	8.82	0.87	1.00	4
Office	8.73	1.01	0.50	6
Recreational	8.55	0.93	1.00	7
Special purpose	8.18	0.60	0.00	8
Storage	8.00	0.89	0.00	9
Agricultural	8.00	1.26	0.50	10
Residential	7.64	1.03	1.50	11
Mixed Use	7.55	1.04	1.00	12
Religious	7.36	0.81	1.50	13

using the appropriate systems for the effective management of facilities. With the wide spectrum of functions delivered by the effective management of facilities, it is germane that some of these functions would effectively encourage the espousal of cyber-physical systems. Due to the differences in these FM functions, some might serve as efficient drivers for the adoption of cyber-physical systems. It is pertinent to discover among the functions of facilities management which are the drivers of the adoption of cyber-physical systems. The study identified 18 FM functions. This list was put to the panel of experts to ascertain which of them would serve as drivers in adopting cyber-physical systems. As shown in Figure 8.4, seven of the facilities management functions were considered to be of very great extent (VHE) (VHE. 9.00–10.00) in the drive for the espousal of cyber-physical systems, while 11 of the functions were deemed to be of high extent (HE) (HE: 7.00–8.99). Conversely, none of the functions was considered not important in the drive for the adoption of cyber-physical systems. Furthermore, the IQD values of the facilities management functions all attained consensus except for two functions: attaining strategic objectives and finance management. Both facilities management functions had an IQD value above the cut-off adopted for this study, i.e. (≤1).

Discussion of results

DSO1 To identify the main features for adopting cyber-physical systems for facilities management

The first objective set for the Delphi study was to identify the main features for the adoption of CPS for FM. The results from the Delphi process indicate

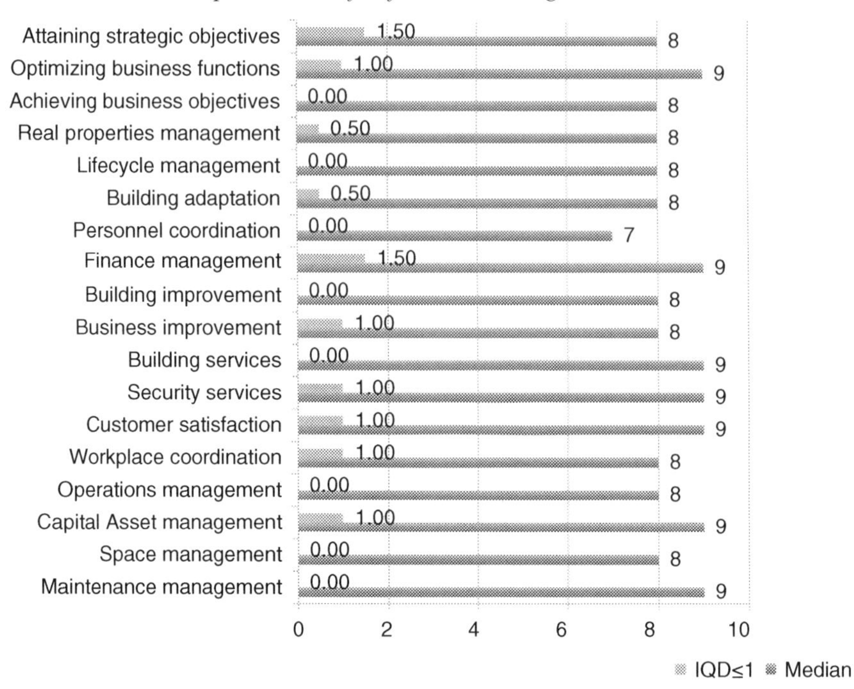

Figure 8.4 Facilities management functions as drivers for the adoption of CPS

that the six main features for the adoption of CPS for FM all attained consensus, hence proving to be significant. This agrees with other studies that have focused on technology adoption (Wallace and Sheetz, 2014). The main attributes all had a very high significant median score except for enabling measures with a highly significant rating. Moreover, the IQD value of the main features all met the cut-off value for attaining consensus. Based on these findings, it is notable that performance expectancies, effort expectancies, enabling measures, social influence, business environment, and performance measurement are key determinants in the discussion on adopting CPS for FM.

DSO2 *To identify the sub-features for adopting cyber-physical systems for facilities management*

The second objective of the Delphi study was to identify the sub-features for the adoption of CPS for FM. Findings from the Delphi process show that the majority of the sub-attributes for adopting CPS for FM are significant. For performance expectancies, of the total 19 sub-attributes, none of the attributes were portrayed as significant, with three of the attributes having a very high significance rating. In contrast, 16 of the attributes had a high significance rating. Moreover, all 19 attributes attained consensus by having IQD values within the ranges of the cut-off value. These findings are corroborated by the various studies which highlighted most of the sub-attributes of performance expectancies to be an important aspect of technology adoption (Boyes, 2013; Eben and Achampong, 2010; Park, 2009; Parn and Edwards, 2019; Watson, 2018).

Furthermore, for social influence with a total of 14 sub-attributes, four appeared to be of very high significance, while ten were indicated to be of high significance. However, two of the sub-attributes (used by better-performing organisations and as status symbols for the organisation) did not attain consensus due to their IQD values exceeding the cut-off. This contrasts with other studies that have shown that these features were important determinants in technology adoption (Davis and Songer, 2008; Talukder and Quazi, 2011). Moreover, sub-attributes such as organisation's support for the use of the system, stakeholders' demand for use, internal network of the organisation, and external network of the organisation all attained a very high consensus with an IQD value of 0.00. This is corroborated by other studies centred on the discussion of technology adoption (Brewer and Gajendram, 2011; Walter, 2006; Wang and Song, 2017).

In addition, enabling measures had 18 sub-attributes, with 6 having very high significance, while 12 had a rating of high significance. However, one of the sub-attributes (interactiveness with the system) did not attain consensus due to its IQD values exceeding the cut-off. This differs from the verdict of other studies, which affirms that the sub-attributes are a major determinant of technology adoption (Lin et al., 2007). Also, sub-attributes, such as compatibility with work procedures, comparison with the previous system, financial resources, and intellectual resource all achieved very high consensus with an IQD value of 0.00. This is supported by Erdoğmus and Esen (2011), who noted that an individual's or organisation's financial capacity is very important in the decision to adopt a particular technology. Furthermore, Talukder and Quazi (2011) emphasised the importance of intellectual resources in the pursuit of new technology since the human resources for engaging such technology concerning intellectual delivery are fundamental to its use.

The assessment of the sub-attributes based on effort expectancies had 15 sub-features. Four of the attributes indicated very high significance, while 11 were seen to be of high significance. Also, all the sub-features attained consensus in the Delphi experts' opinions, with all having IQD values within the cut-off range. The sub-attributes with very high consensus are self-efficacy, the usability of the system, users' willingness to use the system, complexity of the system, and compatibility with the organisation's other systems. This finding agrees with previous studies such as Brinkerhoff (2006) and Moore (2012). They noted that the ability of technology to be user-friendly and its ability to deliver with less input are major deciders in its adoption. Also, the voluntary stance on the part of the user is an important factor in the adoption of a given technology (Lowry, 2012).

Furthermore, the business environment had 16 sub-attributes, with eight being of very high significance, while the other eight sub-attributes appeared to be of high significance. However, three sub-attributes (market orientation, return on investment, and sales growth) did not reach a consensus. This disagrees with the other studies that have portrayed these sub-attributes to be important in the discussion on technology adoption (Hernandez-Ortega et al., 2014; Oh et al., 2014). Moreover, some of the sub-attributes attained a very high consensus, such as the organisation's resource commitment, work procedure support, resource availability, and environmental dynamism. This is in tandem with the study of Aboelmaged (2014) and Lee et al. (2005), who stated that the willingness of the organisation to make a financial commitment by investing in a given technology has a huge influence on the adoption of a particular technology. Hence, the financial responsibility on the part of the organisation is a huge rallying point when technology decisions are to be made. Also, Erdoğmus and Esen (2011) established that the nature of the procedural functions to be engaged in with the intended technology is of vital importance in the decision on its adoption or not.

The assessment of the sub-attributes based on performance measurement indicates that out of a total of 16 sub-attributes, nine were proven to be of very high significance. At the same time, seven were shown to be of high significance. Furthermore, it was revealed that all the sub-attributes achieved consensus based on the cut-off value set for the study. The sub-attributes with very high consensus, having an IQD of 0.00, include the built-in capability of facility adaptation, identification of problems in facilities, improved customer satisfaction, customers' involvement in improving the evaluation process, attainment of customers' specific needs, and timely communication of policy changes. These findings are corroborated by other studies that identified these sub-attributes as significant in the drive for technology adoption (Cheng et al., 2014; Hernandez-Ortega et al., 2014).

DSO3 To determine the extent to which cyber-physical systems can be adopted for the different types of facilities

The third objective of the Delphi study was to determine the extent to which cyber-physical systems can be adopted for the different types of facilities. Findings from the study showed that among the 13 types of facilities identified, the facilities that proved to be best suited to incorporate cyber-physical systems are industrial, healthcare, retail, hospitality, and educational. The reason for such opinions from the experts is not far-fetched. Since cyber-physical systems are one of the conveyors of the Fourth Industrial Revolution, which is based on overhauling production and manufacturing processes, it can be established that the use of such a system in this type of facility is not out of place. Also, it can improve the value proposition exhibited by traditional industry, coupled with the creation of opportunities and business models that are completely new. Data analysis for the improvement of performance is one of the comparative advantages of inculcating a groundbreaking system, such as cyber-physical systems. Also, it was revealed that cyber-physical systems would largely be of benefit in healthcare facilities. This stems from the interoperability of the system with advanced wireless sensor networks and cloud computing. Due to the interaction of computational platforms with communication networks, the overall management of health facilities could be better delivered. Moreover, the experts' verdict that retail facilities would be a good fit for incorporating cyber-physical systems is revealed by the study.

However, findings revealed that not all the highlighted types of facilities attained consensus in the experts' opinions concerning the extent to which cyber-physical systems can be adopted for them. Notable inclusions are residential and religious facilities. The IQD value for these facilities was outwith the range of the cut-off set for the study. This reinforces the stance that the adoption and implementation of a complex system would not be suitable for such types of facilities. Since facilities that serve residential purposes do not necessarily have to be characterised by huge complex systems for their optimum functioning, there would be no need to include a sophisticated system such as a cyber-physical system for its daily use. Likewise, the same can be said for religious facilities, such facilities do not demand advanced technological innovations such as cyber-physical systems.

DSO4 To predict the extent to which the various functions of facilities management can serve as drivers for adopting cyber-physical systems

The last objective for the Delphi study was to predict the extent to which the various functions of facilities management can serve as drivers for the adoption of cyber-physical systems. It has been noted that FM involves a wide spectrum of functions, activities, and operations, often referred to as the scope of FM. However, some of these functions would be able to implement technological innovations more than others. Among the 18 identified facilities management functions, it was shown in the study that seven of the functions are of a very great extent in the drive for the adoption of cyber-physical systems, while 11 functions are of great extent. As revealed in the Delphi study, optimising business functions, building services, security services, and capital asset management are the functions that serve as very high drivers for the espousal of CPS for FM. This is corroborated by Teicholz (2013), who affirmed that incorporating technological innovations into facilities management practices will improve the organisation's business growth and market competition. Furthermore, Love et al. (2013) stated that the espousal of automated approaches in the delivery of facilities management is encouraged by the enhancement in service delivery, reduction of business costs, and effective delivery of maintenance practice.

However, not all the identified functions attained consensus in the experts' opinions in the Delphi study. Notably, functions such as attaining strategic objectives and finance management were not within the threshold of the cut-off value set for the study. This contrasts with the study by Atkin and Bildsten

(2017). Moreover, as revealed in the study's findings, the FM functions that achieved very high consensus are life-cycle management, personnel coordination, building improvement, building services, operations management, space management, and maintenance management. These findings are corroborated by other studies which have affirmed that the attainment of an increase in productivity, and enhanced efficiency in building adaptation and maintenance are some of the inherent benefits of integrating technological innovations in FM practices (Torok, 2014).

Summary

The outcomes of the first and second rounds of the Delphi study were presented in this chapter. For the individual questions posed to the panel of experts, the computations were carried out. These questions dwelt on the significance of the main and sub-attributes for adopting cyber-physical systems for facilities management. Also, the extent to which cyber-physical systems can be adopted for the different types of facilities was determined. Likewise, the prediction of the extent to which the various functions of facilities management can serve as drivers for the adoption of cyber-physical systems was presented. A comprehensive discussion of the study's findings concerning the outlined objectives of the study concluded the chapter. The highlights of the findings showed a coherent discussion among the experts concerning the adoption of CPS for FM by achieving a consensus among the issues of concern. Ultimately, the findings of the Delphi study project the key determinants and constructs in the adoption of CPS for FM.

References

Aboelmaged, M.G. (2014). Predicting e-readiness at firm-level: An analysis of technological organisation and environmental (TOE) effects on e-maintenance readiness in manufacturing firms. *International Journal of Information Management*, 34(5), 639–651.

Addison, T. (2003). E-commerce project development risks: Evidence from a Delphi survey. *International Journal of Information Management*, 23(1), 25–40.

Adler, M. and Ziglio, E. (1996). *Gazing into the Oracle: The Delphi Method and Its Application to Social Policy and Public Health*. London: Jessica Kingsley Publishers.

Aghimien, D., Aigbavboa, C., and Oke, A. (2020). Critical success factors for digital partnering of construction organisations: A Delphi study. *Engineering, Construction and Architectural Management*, 27(10), 3171–3188.

Aigbavboa, C. (2013). An integrated beneficiary centred satisfaction model for publicly funded housing schemes in South Africa. PhD thesis. University of Johannesburg.

Andranovich, G. (1995). *Developing Community Participation and Consensus: The Delphi Technique*. Los Angeles, CA: California State University.

Armstrong, D., Marshall, J.K., Chiba, N., Enns, R., Fallone, C.A., Fass, R., Hollingworth, R., Hunt, R.H., Kahrilas, P.J., Mayrand, S., Moayyedi, P., Paterson, W.G., Sadowski, D., and van Zanten, S.J. (2005). Canadian consensus conference on the management of gastroesophageal reflux disease in adults – update 2004. *Canadian Journal of Gastroenterology and Hepatology*, 19(1), 15–35.

Atkin, B. and Bildsten, L. (2017). A future for facility management. *Construction Innovation*, 17(2), 116–124.

Avella, J. (2016). Delphi panels: Research design, procedures, advantages, and challenges. *International Journal of Doctoral Studies*, 11, 305.

Balasubramanian, R. and Agarwal, D. (2012). Delphi technique: A review. *Journal of Public Health Dentistry*, 3(2), 16–26.

Boyes, H. (2013). Resilience and cyber security of technology in the built environment: IET Standards Technical Briefing. London: The Institution of Engineering and Technology, Available at: https://www.theiet.org/resources/standards/-files/cyber-security.cfm?type=pdf (accessed 4 October 2021).

Brewer, G. and Gajendram, T. (2011). Attitudinal, behavioural, and cultural impacts on e-business use in a project team: A case study. *Journal of Information Technology in Construction (ITcon)*, 16(37), 637–652.

Brill, J., Bishop, M., and Walker, A. (2006). The competencies and characteristics required of an effective project manager: A web-based Delphi study. *Educational Technology Research and Development*, 54(2), 115–140.

Brinkerhoff, J. (2006). Effects of a long-duration, professional development academy on technology skills, computer self-efficacy, and technology integration beliefs and practices. *Journal of Research on Technology in Education*, 39(1), 22–43.

Buckley, C. (1995). Delphi: A methodology for preferences more than predictions. *Library Management*, 16(7), 16–19.

Chan, A.P.C., Yung, E.H.K., Lam, P.T.I., Tam, C.M., and Cheung, S.O. (2001). Application of Delphi method in selection of procurement systems for construction projects. *Construction Management and Economics*, 19(7), 699–718.

Chang, J. and Ding, R. (2022). Theories of journalism in the digital era: Knowledge, value, and conceptual framework. In *Digital Journalism in China*. New York: Routledge, pp. 8–22,

Cheng, L., Wen, D., and Hong-Chang Jiang, H. (2014). The performance excellence model in construction enterprises: An application study with modelling and analysis. *Construction Management & Economics*, 32(11), 1078–1092.

Cuhls, K. (2003). Delphi method. Available at: http://www.unido.org/fileadmin/import/16959_DelphiMethod.pdf (accessed 17 October 2021).

Davis, K.A. and Songer, A.D. (2008). Resistance to IT change in the AEC industry: An individual assessment tool. *Electronic Journal of Information Technology in Construction*, 13, 56–68.

Dunn, W.N. (1994). *Public Policy Analysis: An Introduction*. Englewood Cliffs, NJ: Prentice-Hall.

Eben, A.K. and Achampong, A.K. (2010). Modelling computer usage intentions of tertiary students in a developing country through the technology acceptance model. *International Journal of Education and Development Using Information and Communication Technology*, 6(1), 102–116.

Erdoğmuş, N., and Esen, M. (2011). An investigation of the effects of technology readiness on technology acceptance in e-HRM. *Procedia: Social and Behavioral Sciences*, 24, 487–495.

Ewing, R.P. (1979). The uses of futurist techniques in issues management. *Public Relations Quarterly*, 4, 15–18.

Falzarano, M. and Zipp, G. (2013). Seeking consensus through the use of the Delphi technique in health sciences research. *Journal of Allied Health*, 42(2), 99–105.

Fletcher, A. and Marchildon, G. (2014). Using the Delphi method for qualitative, participatory action research in health leadership. *International Journal of Qualitative Methods*, 13(1), 1–18.

Goodman, C.M. (1987). The Delphi technique: A critique. *Journal of Advanced Nursing*, 12, 729–734.

Hallowell, M. (2008). A formal model of construction safety and health risk management. PhD dissertation. Oregon State University.

Hallowell, M.R. and Gambatese, J.A. (2010). Qualitative research: Application of the Delphi method to CEM research. *Journal of Construction Engineering and Management*, 136(1), 99–107.

Harder, A., Place, N., and Scheer, S. (2010). Towards a competency-based extension education curriculum: A Delphi study. *Journal of Agricultural Education*, 51(3), 44–52.

Hasson, F., Keeney, S., and McKenna, H. (2000). Research guidelines for the Delphi survey technique. *Journal of Advanced Nursing*, 32(4), 1008–1015.

Haynes, C. and Shelton, K. (2018). Delphi method in a digital age: Practical considerations for online Delphi studies. In V. Wang and T. Reid (eds), *Handbook of Research on Innovative Techniques, Trends, and Analysis for Optimized Research Methods*. Hershey, PA: Information Science Reference, pp. 132–151.

Helmer, O. (1977). Problems in futures research: Delphi and causal cross-impact analysis. *Futures*, 9(1), 17–31.

Hemmat, M., Ayatollahi, H., Maleki, M., and Saghafi, F. (2021). Health information technology foresight for Iran: A Delphi study of experts' views to inform future policymaking. *Health Information Management Journal*, 50(1–2),76–87.

Hernandez-Ortega, B., Serrano-Cinca, C., and Gomez-Meneses, F. (2014). The firm's continuance intentions to use inter-organisational CTs: The influence of contingency factors and perception. *Information & Management*, 51(6), 747–761.

Holey, E.A., Feeley, J.L., Dixon, J., and Whittaker, V.J. (2007). An exploration of the use of simple statistics to measure consensus and stability in Delphi studies. *BMC Medical Research Methodology*, 7(52), 1–10.

Howell, S. and Kemp, C. (2005). Defining early number sense: A participatory Australian study. *Journal of Educational Psychology*, 25(5), 555–571.

Iqbal, S. and Pipon-Young, L. (2009): The Delphi method. *Nursing Research*, 46 (2),116–118.

Jones, T. (1980). *Options for the Future: A Comparative Analysis of Policy-Oriented Forecasts*. New York: Praeger.

Kauko, K. and Palmroos, P. (2014). The Delphi method in forecasting financial markets: An experimental study. *International Journal of Forecasting*, 32(2), 313–327.

Kim, J., King, B., and Kim, S. (2022). Developing a slow city tourism evaluation index: A Delphi-AHP review of Cittaslow requirements. *Journal of Sustainable Tourism*, 30(4), 846–874.

Kloser, M. (2014). Identifying a core set of science teaching practices: A Delphi expert panel approach. *Journal of Research in Science Teaching*, 51(9), 1185–1217.

Landeta, J. (2006). Current validity of the Delphi method in social sciences. *Technological Forecasting and Social Change*, 73(5), 467–482.

Lee, H., Lee, Y., and Kwon, D. (2005). The intention to use computerized reservation systems: The moderating effects of organisational support and supplier incentive. *Journal of Business Research*, 58(11), 1552–1561.

Lin, C., Shih, H., and Sher, P. (2007). Integrating technology readiness into technology Acceptance: The TRAM model. *Psychology and Marketing*, 24(7), 641–657.

Lin, Y.C., Cheung, W.F., and Siao, F.C. (2014). Developing mobile 2D barcode/RFID-based maintenance management system. *Automation in Construction*, 37, 110–121.

Love, P.E., Simpson, I., Hill, A. and Standing, C. (2013). From justification to evaluation: building information modelling for asset owners. *Automation in Construction*, 35, 208–216.

Lowry, G. (2012). Modelling user acceptance of building management systems. *Automation in Construction*, 11(6), 695–705.

McKenna, H. (1994). The Delphi technique: A worthwhile research approach for nursing? *Journal of Advanced Nursing*, 19(6), 1221–1225.

McMillan, S., King, M., and Tully, M. (2016). How to use the nominal group and Delphi techniques. *International Journal of Clinical Pharmacy*, 38(3), 655–662.

Moore, T. (2012). Towards an integrated model of IT acceptance in healthcare. *Decision Support Systems*, 53, 507–516.

Niederberger, M. and Spranger, J. (2020). Delphi technique in health sciences: A map. *Frontiers in Public Health*, 8, 457.

Oh, J.C., Yoon, S.J., and Chung, N. (2014). The role of technology readiness in consumers' adoption of mobile internet services between South Korea and China. *International Journal of Mobile Communications*, 12(3), 229–248.

Olij, B., Erasmus, V., Kuiper, J., van Zoest, F., Van Beeck, E., and Polinder, S. (2017). Falls prevention activities among community-dwelling elderly in the Netherlands: A Delphi study. *Injury*, 48(9), 2017–2021.

Park, S Y. (2009). An analysis of the technology acceptance model in understanding university students' behavioural intention to use e-learning. *Educational Technology & Society*, 12(3), 150–162.

Parn, E. and Edwards, D. (2019). Cyber threats confronting the digital built environment: Common data environment vulnerabilities and block chain deterrence. *Engineering, Construction and Architectural Management*, 26(2), 245–266.

Powell, C. (2003). The Delphi technique: Myths and realities. *Methodological Issues in Nursing Research*, 41, 376–382.

Rayens, M.K. and Hahn, E.J. (2000). Building consensus using the policy Delphi method. *Policy, Politics, Nursing Practice*, 1(2), 308–315.

Reid, N.G. (1988). The Delphi technique: its contribution to the evaluation of professional practice. In R. Ellis (ed.), *Professional Competence and Quality Assurance in the Caring Professions.* New York:Chapman and Hall.

Sackman, H. (1974). *Delphi Assessment: Expert Opinion, Forecasting and Group Process*, Santa Monica, CA: Rand Corporation.

Skulmoski, G.J., Hartman, F.T., and Krahn, J. (2007). The Delphi method for graduate research. *Journal of Information Technology Education*, 6, 1–11.

Steurer, J. (2011). The Delphi method: An efficient procedure to generate knowledge. *Skeletal Radiology*, 40(8), 959–961.

Strasser, S., London, L., and Kortenbout, E. (2005). Developing a competence framework and evaluation tool for primary care nurses in South Africa. *Education for Health*, 18(2), 133–144.

Suris, J. and Akre, C. (2015). Key elements for, and indicators of, a successful transition: An international Delphi study. *Journal of Adolescent Health*, 56(6), 612–618.

Talukder, M. and Quazi, A. (2011). The impact of social influence on individuals' adoption of innovation. *Journal of Organisational Computing and Electronic Commerce*, 21(2), 111–135.

Teicholz, P. (2013). *BIM for Facility Managers*, 1st edn. Hoboken, NJ: John Wiley & Sons.

Torok, M. (2014). Image-based automated 3D crack detection for post-disaster building assessment. *Journal of Computing in Civil Engineering*, 28(5), 1–13.

Turoff, M. and Hiltz, S. (1996). Computer-based Delphi processes. In M. Adler and E. Ziglio (eds), *Gazing into the Oracle: The Delphi Technique and Its Application to Social Policy and Public Health*. London: Jessica Kingsley.

Vernon, W. (2009). The Delphi technique: A review. *International Journal of Therapy and Rehabilitation*, 16(2), 69–76.

Wallace, L.G., and Sheetz, S.D. (2014). The adoption of software measures: A technology acceptance model (TAM) perspective. *Information & Management*, 51(2), 249–259.

Walter, M. (2006). Return on interoperability: The new ROI. *CAD USER*, 19(3), 14.

Wang, G. and Song, J., (2017). The relation of perceived benefits and organisational supports to user satisfaction with building information model (BIM) . *Computers in Human Behavior*, 68, 493–500.

Watson, S. (2018). Cyber-security: What will it take for construction to act? Available at: https://www.constructionnews.co.uk/tech/cyber-security-whatwill-it-take-for-construction-to-act-22-01-2018/ (accessed 3 October 2021).

Whitman, N.I. (1990). The committee meeting alternative: Using the Delphi technique. *The Journal of Nursing Administration*, 51(1), 57–68.

Yousuf, M.I. (2007). The Delphi technique. *Essays in Education*, 20, 80–89.

Index

Page numbers in *italics* refer to figures. Page numbers in **bold** refer to tables.

Printed in the United States
by Baker & Taylor Publisher Services